BITES
& STINGS
The World of Venomous Animals

JOHN NICHOL

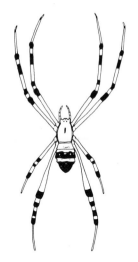

Facts On File

New York • Oxford

Facts on File, Inc.
460 Park Avenue South
New York   NY 10016
USA

Library of Congress Cataloging-in-Publication Data

Nichol, John.
    Bites and stings: the world of venomous animals/John Nichol.
        p.   cm.
    Bibliography: p.194
    Includes index.
    ISBN 0-8160-2233-X
    1. Poisonous animals.   I. Title
QL100.N5 1989
591.6'9—dc20   89–32853

Facts on File books are available at special discounts when
purchased in bulk quantities for businesses, associations,
institutions or sales promotion. Please contact the Special
Sales Department of our New York office at 212/683–2244
(dial 800/K0322–8755 except in NY, AK or HI).

Composition by Typesetters (Birmingham) Ltd,
Smethwick   Warley   West Midlands
Manufactured in Great Britain
by Redwood Burn Ltd   Trowbridge   Wilts

10   9   8   7   6   5   4   3   2   1

Contents

Acknowledgements

The writing of a book is a complicated business, during which the author calls upon the assistance of many people, and can indeed become a pain in the neck to some of them when phone call follows phone call asking for information. It says a lot for the tolerance of humanity that nearly all such requests are met with understanding. At the end of it all, the only way of repaying this generosity is to thank everybody for what they have done. In the case of this book, if I mentioned all my contacts who have helped in so many ways, the volume would be twice as long, so I hope they will forgive me if I say a general and heartfelt thanks to all of them. Some went far beyond the call of duty, and they really must be mentioned by name. Ralph Fitchett and all his colleagues gave me a great deal of time that they could have used elsewhere. Ann Webb, secretary of the British Tarantula Society, was invaluable on the subject of invertebrates, while David Ball in the reptile house at London Zoo was every bit as useful on the subject of reptiles. Most of my fishy information was the result of conversations with Paul Horsman. Chris Miller of Customs and Excise, John Foden of Drayton Manor Park, Tony Gadsby-Peet from Marvel Licensing and Dennis Furnell all helped, each in his own way, to make the book happen.

When it comes to actually producing a volume from a great heap of hand-written scribble, a second category of people is involved. The author gets any credit there may be, but without the photographers, the designers, the typesetters, the proof readers and many, many others, the book would never get on the shelf of your local bookshop. So to all of them and to Margaret for typing the manuscript, which was an awful job, thank you.

Bites and Stings:
The Television Programme

The television programme of the same title that this book is designed to complement, is the result of something that happened some time ago. Each year in the autumn, the British Broadcasting Corporation produces an extremely entertaining, evening-long programme called Children in Need. The whole idea behind this was to raise money for children's charities throughout the British Isles by persuading almost everyone throughout the country to pledge money, or to organise and take part in an amazing variety of stunts, all of them designed to raise cash in one way or another. The programme itself links BBC television and radio stations throughout the United Kingdom, each contributing something to the whole. Last year, the producer of Bites & Stings was approached by a local BBC radio station in the part of the country in which he lives, and asked to arrange some sort of spectacular stunt that would raise a lot of money, and be televisual enough to be incorporated into the television programme the same evening. After rejecting a number of ideas, a national double-glazing company, Anglian Windows, was asked if they would provide a 1.8m (6ft) glass cube with a door in one side. They agreed instantly, and entered into the spirit of the stunt with enthusiasm. A venue had to be found, and the manager of the newly opened local branch of Debenham's department store was happy to agree to the whole operation being staged on his premises.

The idea that had emerged from a number of discussions was that Bill Oddie, the television celebrity, would be sponsored to enter a cage – the glass cube – packed with some thousands of creepy crawlies of all sorts. The project seemed to strike a spark in everyone, and on the morning it was to be staged, long before most people were out of bed, the security staff at the department store were gingerly admitting a number of people via the staff entrance who carried cage after cage bearing intriguing labels like TARANTULAS: THIS WAY UP and JUNGLE TITANS: HANDLE WITH CARE. But the cage that caused most concern was marked SCORPIONS: KEEP WARM.

By seven o'clock the fan heaters in the cage had been turned on, and while waiting for the temperature to rise, the various animal handlers from all over the country sat around and discussed the finer points of tarantula breeding. Half an hour later the stage was being set as animals of all kinds were introduced into the cage. Six large, black, evil-looking Emperor Scorpions were first, followed by stick insects of numerous species, mantises of all sorts, tarantulas, Giant Millipedes, several hundred locusts, beetles in plenty, and a gang of Hissing Cockroaches, looking for all the world like clockwork toys. The heater had worked well, and in the tropical atmosphere the cage was soon buzzing and seething with flying, walking, whizzing and jumping invertebrates. The shop opened to the public at nine, and within seconds the glass cage and the startled organisers were surrounded by a horde of enthusiastic shoppers, a mob which became thicker by the minute until it was impossible to move in the neighbourhood. A class of children from a local primary school had arrived, ready to help push Bill Oddie into the cage. As soon as he turned up Bill was escorted through the throng by the security staff who felt a lot happier now that all the horrors were behind glass. He was gleefully shoved through the door into the cage, and the clock started ticking to measure the time he was in there.

There was not much else to do for the moment, but it soon became apparent that the public was loving the whole event. Bill was quite happy, sitting covered in things from a Transylvanian nightmare while the public crowded as close to the glass as they could, with expressions on their faces that were a mixture of revulsion and shivery delight. As the organisers sat and waited for Bill Oddie to emerge they were approached by many people who wanted to know all sorts of facts about the animals in the cage. Many of the questions were concerned with whether or not they were dangerous, and especially about why they weren't biting, if they were poisonous as some of them were. Other queries were straight natural history enquiries. Where did they come from, was a common question, and what sort of lives did they lead.

When the whole stunt was over, and a profusely sweating Bill Oddie was whisked off to the manager's office for a well-earned drink, it had been anticipated that everyone would drift off and the cage could be dismantled. What actually happened was that the crowds would not move. They stayed to gasp at the creepy crawlies, so in order to continue the fund- raising aspect of the stunt, first one of the animal handlers and then another entered the cage to spend some time sitting quietly while beasts of all kinds clambered over them. The audience loved it. The piece went out that evening during the television programme, and since then it has proved to be one of the most popular segments. Whatever parts of the programme were watched by viewers, it seems to have been that bit that stuck in their minds, and Bill Oddie was saying recently that ever since that night people have been stopping him to ask questions about that stunt and to comment on his bravery.

In reality there was no risk, even from the comparatively small numbers of venomous animals that were in the cage, but the audience clearly felt that there was, or if there wasn't that there ought to be. The horrified fascination on the faces of the people who were watching stimulated a discussion on why man has this peculiar love/hate relationship with venomous animals. Had Bill Oddie been in the same cage with half a dozen baby lions, the reaction would have been quite different. Instead of ugh! it would have been ahh!

Once the idea of a programme about man's relationship with venomous animals had emerged, research showed how intricate such a concept is. As we will see in this book, cultures all over the world and throughout history have had an involvement with venomous animals that was far greater than if they had not been venomous, but a way of telling the story for television had to be found because an awful lot of potential viewers find the idea of snakes and spiders, bees and jellyfish so absolutely revolting that whenever such topics are shown on television they turn over to the James Bond film on another channel. Many people are very uneasy about animals that they find threatening.

Ultimately someone had the brainwave that a programme made in the manner of a spoof gothic horror movie, containing plenty of humour, ought to work, and at the same time de-fuse the tension. This idea had the bonus that it ought to attract viewers who might want to watch for a variety of reasons other than because they were interested in the animals themselves. Once the idea had been planted it soon emerged that the concept of a comic treatment was brilliant, as the topic lent itself admirably to the introduction of jokes, and the first of these was a splendid suggestion by one of the production team that the opening credits should be rolled over the music of the old song 'I've Got You Under My Skin'.

From the beginning, everyone involved in the project was enthusiastic, adding points, songs and gags all over the place, and before long all of them were on the lookout for venomous animals to include in the programme. Once we started looking, these were to be found everywhere, and as many as possible have been included in the programme. But inevitably, if the programme was to appeal to as wide an audience as possible, much serious matter would have to be left out. The idea of the programme was to persuade viewers to look at venomous animals with new eyes and, it was hoped that, as a result, some of them would realise that these creatures were not as threatening as had been thought previously. At the same time, the programme might stimulate some people to want to find out more about the topic. Interestingly we soon found that, although there have been plenty of programmes and publications about venomous snakes, or venomous spiders, or venomous marine animals, the subject had not been tackled as a whole before. The books that had been published in the past, with a few exceptions, also tended to be technical works, or aimed at readers with an interest in the topic already. We hope that this programme and book will therefore be looked at by the layman as well as the specialist.

Wading through the literature on the subject, one soon discovers that much information that ought to be available to the general public is not easy to come by. Nowadays more and more of us are travelling to ever more exotic parts of the world for our holidays, where we are far more likely to meet snakes or scorpions than in Bognor or Bangor, and there is almost no clear information for such people on what to look out for, and what to do if they are bitten or stung by a venomous animal. In addition, exotic pets like spiders are being kept far more commonly than a few years ago, and every one of the enthusiasts who cares for such animals wants to know more about them, what to do if one of them bites, and where to go for help in the event of such an accident. Researchers on the book and the programme soon found themselves going round in circles when they tried to find out from the medical profession what should be done in the event of an injury from a venomous animal. There are clear answers, but whether you find them or not in an emergency depends on who your doctor is. You can waste valuable time while he discovers that he should be phoning the Poisons Unit and tries to locate the nearest source of an appropriate antivenom. Some doctors are great and know exactly where to look in such an emergency; but for others, for their patients who keep venomous animals, for the holidaymaker in exotic places and for the general reader who apparently represents most of us in finding the whole business of venomous animals absorbing, this book ought to provide many answers.

Once you start to think about how absorbed we are by venomous animals, you soon discover that weekly, literally weekly, there are newspaper reports about bees or scorpions or something venomous, and only recently a cutting arrived from a Far Eastern newspaper with the riveting headline, 'Doctors Fish Scorpion Out Of Man's Ear'. It tells of an unfortunate chap in Saudi Arabia who went to his doctor complaining of earache. The doctor found that a scorpion had taken up residence therein, and the man was admitted to hospital to have it removed. It sat quietly in this warm, dark, humid cave quite happily for ten hours, and it was only when the doctor began to haul it out into the daylight that the scorpion stung the patient.

The programme 'Bites & Stings' is a fun programme. This book gives the opportunity to put much more information before those who want to find out more, but from the very beginning the producers were adamant that all the informative parts of the programme should be scrupulously factual.

The filming of the programme fell into two parts. In the first instance the crew had to travel to many countries around the world to shoot pictures of snake charmers, of rattlesnakes in the wild, of Blue Ringed Octopuses in the waters off the Australian coast, and of laboratories and snake farms involved in the preparation of antivenoms. This location shooting proved to be highly interesting on occasion, such as when a bored King Cobra decided that a life of show business really was not for him, and took off in the direction of the cameraman's feet. Cameramen being the phlegmatic characters they are, this one merely raised his eye from the viewfinder and enquired casually of the producer as the snake eased its way between his feet, 'I suppose I am safe?' and on being reassured, returned contentedly to his work. In other places care had to be taken each morning to ensure that no scorpions had taken up residence inside a nice warm shoe during the night. For some reason one particular member of the crew seemed to attract these little animals. On the first morning when he tapped his shoe and a scorpion fell out, he leapt with alacrity onto the bed. By the end of a week he didn't even bother to move his feet as the arachnids dropped onto the floor, but his attraction for scorpions caused the animal man on the crew to complain bitterly that no scorpions came to take up residence in his shoes, and that some people had all the luck.

This part of the filming was at times hot and uncomfortable, frequently frustrating, but in the end always interesting. The second part of the filming involved the studio work. Contrary to common belief, studio programme making can be extremely tedious, but this one proved to be a delight from the first day, when the actors who were to star in the programme were flown in from the United States or driven to the studios from different parts of Britain. The sets had been built, the costumes had been made, and early one autumn morning the filming began. The story goes that when Vincent Price was making the horror movie *The Fly*, years ago, parts of the script were so funny, or else just plain ludicrous, that in some scenes he and the other actors had to play back to back to prevent the inevitable gales of laughter that would result if they faced each other. On occasion the same sort of atmosphere prevailed here, and throughout the shoot small details and little visual, or sometimes verbal, jokes were constantly added. Making a television programme is a highly professional, expensive business, and some marvellous ideas simply had to be dropped due to the various pressures on the production. The shooting went without a hitch, well, almost without a hitch, though there were a few hairy moments such as when someone left a box of tarantulas on the floor without a lid so that when the studio lights warmed them up they decided to go for a tour of exploration, and some of them had to be rescued from the most unlikely hidey-holes beneath cameras and in corners.

The editing of any programme leaves much of the original film on the floor of the cutting room or, more precisely, in the bins that are there to stop the delicate filmstock becoming damaged on the floor; but even so when the programme was finished it was packed full of so much detail that it is worth watching more than once to try and spot all that is happening.

One aspect of the filming that proved to be more difficult than had been anticipated was the dearth of venomous snakes available in this

The result of a bite by a Loxosceles spider two months after the incident (Herbert Lieske)

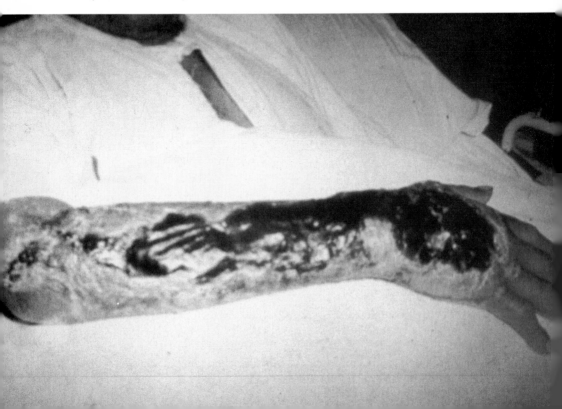

country in captivity. At one time not many years ago, plenty of zoos maintained collections of such animals, and indeed private collectors who kept venomous snakes were far from uncommon. The latter practice has all but stopped thanks to the introduction of the Dangerous Wild Animals Act. Everyone on the production knew this, but what surprised us was just how few zoos keep venomous snakes these days. With all the care in the world there is always a chance of a very nasty accident, and nowadays the risk is difficult to justify. Those collections that still exist are maintained by dedicated people in conditions of the very highest security, but more than one curator pointed out that the collections were being run down and that as the remaining animals die they are not being replaced. One cannot blame anybody for this policy, but it is sad, for however often one watches television programmes about venomous snakes there is nothing quite like a real, living reptile motionlessly watching one through the glass of a tank to make one realise the danger that these animals can represent in the countries in which they originate. Seeing a large Black and White Necked Cobra like the specimen at London Zoo brings home, like no television pictures ever can, just how all the myths and legends concerning snakes came into being many centuries ago when early man first came into contact with these animals that are capable of causing silent, agonising violent death.

Introduction

It seems that nobody is indifferent to venomous animals, though who you are and where you are in the world will affect the way you have been brought up to regard them. In the tropics where such animals are most common, they are generally regarded as either a nuisance or an exploitable resource, while in the developed Western countries it is not uncommon to find even the most rational of souls flapping about in a panic when a wasp appears over the horizon to disturb a summer picnic. Such a reaction is odd since many people go through life without ever once receiving a sting from one of these animals, while the chance of suffering serious injury or death from snakebite is tiny. Somebody has calculated that death from snakebite in the United States is less likely than being struck by lightning, while in Britain it must be infinitely less since there is only a single species of venomous snake, and deaths from its bite are extremely rare despite the fear and hatred that it engenders. Such an attitude is clearly not innate since it is frequently observed that a child brought up in an environment where such feelings are lacking, does not grow to fear or hate animals. It is only when the child goes to school that he or she learns that one is not supposed to feel affection for them.

One of the commonest misconceptions about venomous animals is that because they are capable of inflicting pain or death they spend their lives looking for humans to attack. The reality is that, like most other species, they would far rather keep away from us than do us any harm. It is probably true to say that virtually all injuries and deaths that are attributed to venomous animals are directly related to interference by man in the first place, even though such interference may be unintended. It is certainly the case that if one meets a snake or a scorpion in the wild and walks past it, or even sits down to watch it from a few feet away, it is not going to come across to give the watcher a quick bite before continuing with its own business. Most injuries to people are caused by handling wild animals that not unexpectedly take offence at this sort of treatment, and attack. Most other incidents happen through a person inadvertently disturbing an animal. Pushing a hand into a woodpile to remove a log beneath which is a sleeping rattlesnake is likely to provoke a response, as is walking through an agricultural area in the Far East without any shoes. The results are predictable, and numbers of people are bitten by venomous snakes that have been disturbed in this fashion.

Given that venomous animals can and do cause pain and death, it is not surprising that people come to dislike them, and frequently kill them as a safety precaution. The sad thing is that a great number of perfectly harmless animals are also destroyed when man fails to differentiate between venomous and harmless species. This universal fear becomes far more interesting when one discovers the extraordinary attitudes that have developed in different parts of the world. As a result in many cultures various species of venomous animals have come to be venerated, and in Western society it is common to find that they are regarded with exactly the same sort of horrible fascination that makes people climb onto fairground rides to enjoy the queasy sensation of being flung upside down at high speed for a couple of minutes. This form of appreciation of venomous animals has resulted in a whole variety of entertainments featuring cobras, 'giant tarantulas', 'man-eating scorpions' and much else.

This book looks at venomous animals from all points of view. We start with their natural history. As animals they are absolutely fascinating. Many have developed beautiful, ornate forms, many are brightly coloured and even a superficial look at them can convert someone with a hatred of all creepy crawlies to become an interested observer. They have all evolved with venom apparatus as a means of capturing food or for defence or both. In a world packed with competitors each animal has to exploit its own environment as efficiently as possible so while some, like the giraffe, have developed a means of reaching the parts of a tree that others cannot reach, and some, like a frog-eating bat in Central America, have learnt to live on a highly specialised diet, most venomous animals use venom to obtain food in the most efficient way possible. Some of them use the same equipment to defend themselves by biting or stinging potential attackers. Others have developed a method of protecting themselves from being attacked, again using venom, but in a very different fashion. These animals do not have any apparatus to inject the toxins into an attacker, but rather they store it in various body tissues so that when they are eaten the attacker suffers anything from mild discomfort to death. They are generally referred to as poisonous rather than venomous.

Just as with the latter, man's relationship with poisonous species is complex and interesting. Puffer fish, for example, are beautiful little marine fish that can be found throughout the seas of the tropics. They have an ability to puff themselves up to several times their normal girth when they are threatened, often until they are almost spherical. They are poisonous to eat, yet in Japan where they are referred to as *fugu*, the meat of the puffer fish is regarded as a delicacy. The preparation of this dish is a long, complicated affair that cannot be hurried if one is to remove the toxins from it, and fugu chefs have to undergo a long and rigorous training. Even when they have learnt the techniques they are not allowed to practise until they are licensed. Diners who are determined to eat fugu, if they have any sense, go to restaurants with a reputation to preserve, and which are known to employ only licensed chefs. However as in every business, there are cowboy outfits, and unlicensed fugu chefs are to be found throughout the big cities. They have no shortage of customers either, it would seem, since every year sees reports of a number of deaths from fugu poisoning.

We will also look at venom, how it works, and what can be done to treat injuries caused by venomous animals. The world is full of stories

Pop singers sometimes take the name of venomous animals. Adam Ant was probably the most flamboyant (Richard Young/Rex Features)

about suitable treatment for snakebite or scorpion stings. Most of these are either completely useless or positively harmful, and the only really effective way of treating these conditions is to use antivenom. Finally there is the relationship that man has developed with venomous animals, from the snake charming of India to the rattlesnake cults of the south-eastern states of America, to the feature films that attract viewers in droves. We examine all these and a host of others, and hope to show that though venomous animals should be treated with respect, a snake is no more revolting than a lion, nor a Gila Monster any more scary than a Slow Worm.

The first thing we need to do is to establish what a venomous animal is. For the purposes of this book a venomous animal is one which is capable of producing venomous substances as a means of defence or as a way of obtaining food. It is difficult to be more specific than that since the first thing we learn about the subject is that there are no easy answers. Not all these venomous substances are lethal to man, though the book will concentrate on those that are. Apart from the birds, most of the major groups of animals have venomous members.

Since this is a book about man's relationship with the venomous animals with which he shares the world, it will concentrate more on those species that have most impact on our lives. It is not a book for the specialist, but rather for the person who knows little about the subject but is ready to enjoy a programme on television whether the programme is about the natural history of the cobra, or a horror movie entitled *Return of the Tarantula*, or even Spiderman. Every television producer in the world knows that as soon as a programme about snakes goes out on air there will be a string of telephone calls from irate viewers objecting to this sort of material appearing on our screens. If some of those people were also to read this book, and hopefully thereafter look at venomous animals in a more sympathetic light, they would discover that these animals are not the slimy, horrible monsters of fiction.

Animals of all sorts are under threat today for a variety of reasons, and though an individual person may detest venomous animals, they play an important part in the world, and ultimately our lives in one way or another. In Thailand, cobras are collected in colossal numbers for export to Chinese communities around the world – but especially in the Far East where it is believed that the blood and the gall from a cobra is efficacious in the treatment of a whole variety of disorders – and to provide the skins for the manufacture of so-called luxury leather goods. In Bangkok alone there are several hundred shops that sell cobra-skin products. The result that this destruction of snakes is having on the countryside is considerable. In agricultural areas the population of rodent pests has increased considerably over the last few years to an extent where nowadays many farmers are having real problems obtaining a reasonable yield from their planting. Not one of them has any doubt that the reason for the increase is the disappearance of cobras, which acted as a major control.

These peasants find themselves in a real quandary. Being opportunists they have always made extra money where they could. One way of doing this was to catch any wild animals to sell to the agents of the major animal dealers in the country, and while the small boys spent considerable effort removing the chicks of parrots and mynahs from their nests each spring, the men of the village would catch any cobras that they happened to encounter during the day and take them home to keep in huge concrete containers beneath their houses until the next visit by the dealer's agent. This of course led to their present problem, but when a few years ago cobras seemed abundant the villagers became used to the increased standard of living that the capture of the cobras ensured. Nowadays with less snakes to catch they are poorer, and though they know that each additional reptile captured worsens their problem they cannot afford to ignore any they find and the more the crop is destroyed, the more they need to look elsewhere to supplement it.

The same sort of story is being repeated throughout the world. If for no other reason, we cannot afford to destroy any species of plant or animal because it is a nuisance or because we wish to use the body for something. At one time conservationists warned that destroying any species could lead to unforeseen consequences. Nowadays we know only too well what the consequences will be, and though the Thai peasants who catch huge numbers of cobras are seriously affecting their survival the major threat to populations of all sorts of animals throughout the world is the destruction of habitat. That term has been bandied about so much over the last few years that it tends to provoke a yawn whenever it is heard, and it is only when one stands on a hilltop in the tropics, ankle deep in black ash, and looks across a valley covered in the same depressing ash, and learns that only a week before this was all rainforest, that one understands its meaning. Each acre of rainforest contains millions of, say, spiders. Add to these the millions of other creepy crawlies, the birds, the mammals, the frogs, and everything else that lives there, and it soon becomes apparent how much is destroyed when a forest is logged and burnt. Most countries already have legislation to stop this happening but there is too much money at stake for anyone to take much notice, and indeed frequently the very government departments that are supposed to prevent this destruction are involved in it.

The reason for the logging is easy to see. One only has to look through the local newspaper to find newspaper advertisements for teak coffee tables, mahogany framed conservatories and rosewood ornaments. Virtually none of these timbers come from plantations. They are nearly all hacked out of the rainforest. Not only are the trees cut down for their timber, they are dragged out through the undergrowth destroying much more than just the trees that are required. And don't let anyone tell you that there is a positive side to all this as the local villagers earn money from helping in these operations. Perhaps at one time they did, but nowadays the only people to get rich are the syndicates in the major cities and their shareholders around the world. It is they who send in huge vehicles that cut down and remove the trees quickly and efficiently and then move out again. The locals are left to contemplate the mess. After

Toads have long been regarded as witches' familiars. Some of them were princes that had been put under a spell which could only be released when the toad was kissed by a beautiful princess (John Nichol)

a few days they set fire to all the dead and dying timber that is left and plant a crop on the soil and ash that remains. That crop will probably do well, but come the first rains and all the topsoil washes away down the streams to the rivers and disappears forever. Next year there might well be floods in the region since there are no trees and other plants to soak up all the water. Conservation is important to all of us, even the conservation of venomous animals.

A NOTE ON WORDING

The layman normally talks about 'poisonous animals' whereas the professional who has to do with these beasts refers to them as 'venomous animals'. Use which you wish, either will do, but in this book the word 'venomous' refers to animals that are capable of ejecting venom in one way or another, at or into their prey and their enemies. 'Poison' is used to refer to animals whose meat or other tissues are poisonous, and which causes sickness or death in any animal that eats them. A cobra is venomous; a puffer fish is poisonous.

To make things easier for readers who wish to read further on a particular subject, common names of animals are spelt with capital initials, but groups of animals are not, so we can say cobras are venomous and puffer fish are poisonous but Black- and White-lipped Cobras are venomous, and the Spotted Porcupine Fish, which is a species of puffer fish, is poisonous.

1 · The Natural History of Venomous Animals

Venomous animals of one sort or another are to be found pretty well all over the globe with the exception of the very coldest parts, though the number of species is greatest in the tropics. There are various animals that appear at first glance to be venomous, but which are not covered by this book as closer examination will reveal that they inject other substances than venom. Diseases such as malaria, rabies and tripanosomiasis are all transmitted via the bites of animals but are not due to venom. They are the result of infection by one or another micro-organism that is injected into the victim via the saliva of the attacker. The extent of animals that are actually venomous is surprising as can be seen from the following examples:

Mammals	Duck-billed Platypus
	Water Shrew
Reptiles	Cobras, coral snakes, sea snakes, kraits, mambas, vipers, pit vipers, the boomslang, and a whole variety of less venomous back-fanged snakes.
	Gila Monster, Beaded Lizard
Amphibians	Salamanders
	Newts
	Frogs
	Toads
Fish	Weever fish
	Stingrays
	Lion fish and Scorpion Fish
	Catfish
Arachnids	Spiders
	Scorpions
Insects	Ants
	Bees

	Wasps
Echinoderms	Sea urchins
	Starfish
Coelenterates	Jellyfish
	Sea anemones
	Corals
Molluscs	Cone Shells
	Octopuses

There are also a few other odds and ends such as *Hydra* and some millipedes and caterpillars, but since their venoms are of no significance to man they are mentioned only in passing.

The whole subject of animal venom is full of oddities, and much of it is completely incomprehensible, such as the fact that some tiny little beasts may have enough potential to kill several dozen human beings with each injection, or some thousands of mice. Such a capability is clearly unnecessary to the animal, and this will be examined later. Certainly most of these animals are quite happy to go about their daily lives if you leave them alone and virtually all of them will attempt to escape in the first instance if they possibly can. When this doesn't work they will frequently try bluff to deter an aggressor, puffing themselves up, spreading a hood or showing flash colours to frighten away the threat. If all this fails they will often strike in such a way that they do no harm, often with their mouth closed. Only when they really have no options left will they actually strike to hurt. There are exceptions to this. A jellyfish will drift along without doing any of these things because they are not in the nature of jellyfish, and Puff Adders are great believers in a policy of strike first and ask questions afterwards; but as a generalisation it holds true. As long as one is sensitive to an animal's needs there is no reason at all why someone who is interested in its behaviour should come to harm.

MAMMALS

WATER SHREWS

Neomys fodiens, often erroneously called Water Rats, are carnivorous mammals with a wide distribution from Britain in the west, eastwards to the Pacific Ocean. They are one of the largest shrews, living as their name implies, along the banks of rivers, streams and lakes. Their saliva is venomous to enable them to paralyse the fish and frogs that form their diet. There is evidence that some, if not all, shrews are similarly venomous but to a lesser degree, but none of them is any threat to man.

DUCK-BILLED PLATYPUS

Ornithorhynchus anatinus, that oddest of animals from a country noted for its strange fauna, is one of the native Australian marsupials. A platypus looks like a brown king-sized mole that has decided to go to a fancy dress party as Donald Duck. They have spurs on their front limbs that can be jabbed painfully into an attacker, and through which venom can be injected, but since platypuses are secretive, rare animals that few humans are privileged to see, they can hardly be thought of as dangerous.

REPTILES

SNAKES

There are about 3,000 species of snake in the world, but contrary to popular belief only about 10 per cent of these are venomous. Nonetheless, in many places every snake is regarded as such and destroyed on sight. People who are bitten by non-venomous species sometimes appear to exhibit symptoms of snakebite, but these symptoms are usually a result of mild secondary infections, or indeed as a result of a psychosomatic response on the part of the victim. Some venomous snakes are terrestrial, some arboreal, some spend part or all their lives in freshwater, and the sea snake can be every bit as marine an animal as a fish. Snakes are to be found as far north as Norway and south to Argentina, but despite this wide distribution there are a number of islands around the world where these animals are not to be found.

Some snakes, such as the cobras and their relatives, lay eggs, whilst others, the vipers for example, give birth to live young; but with a subject like this one has to qualify many statements as there are exceptions everywhere. Just to make life difficult, some species of snake are ovoviviparous, meaning that a fertilised egg develops inside the mother until it is ready to hatch, at which point the baby emerges from the egg and from the mother, appearing to be born alive. Because of their shape snakes are viewed with suspicion throughout the world, but in fact they are an animal like the rest of us, which is to say that they possess lungs and hearts and all the other essential bits of anatomy, and therefore mate in pretty much the same way as higher animals, though the form of a male snake's penis is very different. Newly born venomous snakes are every bit as dangerous as adults, and youngsters are often far more aggressive than their parents. Baby cobras are really fierce little beasts, and will even strike while they are actually emerging from the egg.

Venomous snakes vary considerably in size from small species like some of the desert vipers, to the magnificent King Cobra that can grow to around 4.9m (16ft). One expects cobras when they rear and spread their hoods to stand only a foot or so high, but the first time one comes across an angry King Cobra in the wild with his head level with one's chest can be a pretty chastening experience even though they are such beautiful animals.

Ever since the development of modern medicine and probably long before that, man has attempted to perfect a system of identifying immediately whether or not a snake is venomous. There are all sorts of ways of doing this including counting the rows of scales and other equally impractical suggestions, but the real truth is that there is no way of telling whether or not a snake is venomous just by looking at it. Interestingly enough a friend told a story of his time as a national serviceman in the Far East. He was a keen herpetologist and used to keep a variety of snakes in his barrack block. At the time he was due to go on leave his unit was about to depart for jungle manoeuvres, and his companions offered to bring back any snakes they found and keep them until his return. 'But not any venomous ones', he warned them, and went off for his holiday. On his return he was delighted to find that there were quite a number of new specimens awaiting him in a box. After examining them he re-marked that he was impressed that not one was venomous. He was told that there were some but that they had been placed in another, more secure container, and when he came to examine it he was astonished to discover that indeed all the snakes within were capable of inflicting a venomous bite. He asked his friends, none of whom knew the first thing about snakes, how they had identified them. Their reply was 'They just looked poisonous!'.

Since there is no easy, immediate way of distinguishing venomous snakes from non-venomous ones, considerable confusion frequently fol-lows the admission to hospital in different parts of the world, of patients suffering from snakebite. Even if a snake is venomous, there are different types of venom and it is absolutely useless to treat a victim with the antivenom for cobra bites if the injury was caused by a viper. Just to confuse the issue further, antivenom obtained from Indian specimens of a particular species will not prove of much use against the bites of Thai cobras even though they may be of the same species, and the same is true of the different populations of venomous snakes around the world.

From the point of view of this book one can conveniently divide snakes into four groups. Firstly there are the non-venomous (aglyphous) snakes. About 2,500 species of snakes belong to the family Colubridae, most of whose members are non-venomous. Some are, however, and these have poison fangs at the rear of the mouth. Snake venom is really a highly specialised saliva, and for these rear-fanged (opistoglyphous) snakes to do any harm, the animal being bitten must be held in the mouth long enough for the venom to trickle down the grooves of the fangs and into the injured tissue of the victim. With the exception of a couple of species, the venom of back-fanged colubrids is not very dangerous to man. The most venomous of these snakes is the Boomslang which was thought not to be dangerous until a comparatively few years ago when a South African herpetologist of renown proved conclusively that it could indeed kill a human being when he was bitten. The third group of venomous snakes are the front-fanged (proteroglyphous) cobras and their relatives. This family, the Elapidae, have front fangs like hypodermic syringe needles through which they inject venom. They are responsible for a number of deaths each year. A very few species have developed a novel method of defence. Rather than inject venom at predators they force it out through the openings in their fangs, under considerable pressure, aimed directly at the eyes. These snakes are known as Spitting Cobras. The final group are the solenoglyphous snakes. These are front-fanged vipers whose teeth work in the same fashion as those of the cobras except that, whereas those of the Elapidae are fixed in position all the time, those of the vipers are swung back out of the way when not in use. Venom fangs are delicate items of equipment and are easily broken or pulled out, which is why both these front-fanged groups have spare fangs constantly growing behind those actually in use, so that whenever anything happens to an existing fang, the next one takes its place.

As a snake grows it needs to change its skin periodically. It usually gets rid of the old skin in one piece by rubbing its nose against something suitably rough to start it coming off, and once it is torn the snake squirms and pushes against the ground or whatever is available until slowly the whole lot peels off inside out, complete with the transparent scales that had covered the eyes. For a few days prior to sloughing, which is what this operation is called, as the eye scales loosen the snake finds it rather difficult to see and it will often become more ready to bite because of the insecurity of dimmed vision. At this time the eyes appear a cloudy grey colour until the old skin is removed.

All snakes are carnivorous. When a venomous snake is hungry it either goes out looking for a meal, as in the case of the cobras, or it sits and waits until something edible comes within range, which is typical viperid behaviour. Each type, however, uses the other method on occasion. Snakes identify food by sight, though some see better than others, and frequently a snake will not see a prey animal unless it moves. It also uses its tongue to pick up scent particles from the air which it then transfers to the Jacobson's organ inside the mouth where it is able to identify the nature of the various smells. Heat-sensitive pits on either side of the face of the pit vipers and some others enable them to locate the direction and distance of a prey animal very accurately. These heat-sensitive pits are so efficient that experiments have demonstrated that they alone can locate prey even when a snake is unable to use other senses. Finally the snakes make use of what passes for hearing. No snake can hear, but all are very sensitive to vibrations passing through the ground to the body. It is a still perpetuated myth that the cobras used by snake charmers can hear the music of the flute. What actually happens is that when the lid of the cobra's basket is removed in full sunlight, the poor old snake jumps up wondering what has happened. It cannot see very well in the glare; all it is aware of is that there is something waving about in front of its face that is close enough to represent a threat, and in the confusion it follows this about in an attempt to make the necessary calculations for a strike. It so happens that the snake charmer is an excellent snake psychologist and he can judge the speed, the distance and the movement of the moving object – the flute – just right so that the snake appears to the audience to be dancing in front of it. It is purely coincidental that there is music coming from the end of the instrument.

When a venomous snake has located its prey and is within range, it strikes. The speed of a strike varies enormously, but generally the vipers strike far more quickly than other snakes, and certainly if one is alert one can dodge a cobra strike, though one is not encouraged to try. A viper will usually give a quick bite and then sit back and wait for the venom to take effect. When it judges that the victim has succumbed, the snake will set off in pursuit. Elapids on the other hand tend to hang on for a bit, and back-fanged snakes certainly do, munching away at the same time to ensure that venom enters the bloodstream of the animal as quickly as possible since its apparatus is not as efficient as that of the front-fanged snakes. When the prey is dead, or virtually dead, the snake will find the head and start to swallow at that end. This is most efficient because as the animal goes down, all the limbs naturally fold flat along the body, making it easier to swallow. Sometimes a snake will get it wrong and try to swallow an animal from the other end, and provided it is not too big the snake will generally succeed, though much more work is involved. A snake is unable to bite chunks out of an animal, and must always swallow it whole. They are quite amazing creatures in that they can ingest animals several times larger than their own head without any trouble at all.

When a snake starts to swallow a meal it unhinges both jaws at the back so that the mouth can open far wider than appears possible when it is in repose. Once the prey is inside the mouth, the snake can move one half of its lower jaw forward to pull the animal further in and hold it there with the teeth on that side while the other half of the lower jaw performs a similar movement. Consuming a meal can take quite a while and when all of it has disappeared the snake spends some minutes moving the head and neck from side to side, and stretching, to force the food further down the gut. When it is comfortable the reptile will wander off to find somewhere quiet to digest the meal in peace, a process which can take days or even weeks depending on the size of the meal. At this time the animal is vulnerable to attack, and if it becomes necessary for it to fight or flee, a snake can disgorge a partly digested meal in a few moments. If this happens and then the threat goes away the snake may swallow the meal a second time, though this is relatively uncommon. It is this behaviour that must have given rise to the myth that a snake will cover its prey with saliva before swallowing. Regurgitated prey is certainly coated in digestive juices when it comes up.

Most snakes in the wild will only take prey that they have killed themselves though there are exceptions to this, and one or two species, notably one of the Indian water snakes, *Natrix piscator*, will eat anything with animal origins. The books say that it lives on fish and frogs, but it is most certainly not averse to carrion, or even any odd lumps of meat it happens to come across. To watch a snake feed for the first time is an amazing experience, but beyond what is described above there is nothing magical about the feeding habits of these reptiles. Tales such as a snake's ability to mesmerise prey so that it sits still, waiting to be eaten, are only tales. The story persists, however, and was shown in the cartoon film *Jungle Book*, when Kaa, the python, attempted to hypnotise Mowgli.

A venomous snake is able to inject venom repeatedly without having to wait until the glands are recharged, though it normally would not need to do this when hunting, only perhaps as a defence measure. Furthermore, there is strong evidence that a snake is able to adjust the amount of venom that it injects into a prey animal depending on the size of the meal. If the animal is a big, strong one, the snake will automatically use more venom than for a small one.

Snakes do not need to feed every day, and though the period between feeds can vary from a few days to several weeks, the really large animals can afford to go for many months without food if they really need to. All reptiles are cold blooded, which does not mean that their blood is necessarily cold but that they have no mechanism for maintaining it at a steady temperature as we do. Instead their temperature is maintained at the same level as their surroundings. Because of this, snakes are not very efficient and active if they are cold. For this reason the first thing a snake does each morning is find a sunny spot for itself as early as it can, and there it will lie, soaking up heat until it is ready to begin its day. But while a lot of snakes are active during the daytime, others are nocturnal, and these too must find suitable places where they can warm themselves. Consequently in some parts of the world it is not uncommon to find snakes resting on roads during the early part of the hours of darkness since the asphalt retains the heat.

Another distinctive feature of snakes is that they have no eyelids and this makes it difficult to decide when a snake is sleeping. Many snakes remain astonishingly aware of what is happening in their immediate surroundings even when they are asleep, but sometimes it is possible to approach a sleeping snake very closely indeed before it wakes with a start. The lack of eyelids means that if one is not sure whether a legless reptile is a snake or a lizard, one can watch the eyes of the animal to see if it blinks, which will help to identify it. Another distinctive feature is the forked tongue of the snake, for while lizards' tongues come in a variety of forms, those of snakes are always forked, the reason being that each part of the fork is inserted into the paired Jacobson's organs inside the mouth. The tongue, incidentally, is no more harmful than yours despite the persistent stories that insist that it is a stinging organ.

Back-fanged Colubrids Most typical snakes are colubrids and, as we have seen, most of them are non-venomous. The sub-family Boiginae, however, contains a number of back-fanged snakes which have between one and three fangs at the rear of each side of the top jaw adapted for the delivery of venom by being grooved down the front edge. The African sand snakes, *Psammophis*, have a group of enlarged teeth in the middle of each side of the upper jaw in front of the venom fangs. As with most colubrids, these snakes tend to be long, slender and lithe. They inhabit a variety of habitats from deserts to mangrove swamps. There are three back-fanged species in Europe, the Montpelier Snake, *Malpolon monspessulana*, the Hooded Snake, *Macropotodon cucullatus*, and the European Cat Snake, *Telescopus fallax*. Apart from those and a comparatively small number of viperids, the rest of the serpent fauna of Europe is non-venomous. By way of an interesting contrast, two-fifths of the snakes of the island of Madagascar are back-fanged colubrids, and in Australia there are more venomous species than non-venomous – a situation which is unique.

Asia generally is rich in back fangs, species of *Boiga* being commonly encountered throughout the Far East. Undoubtedly the best known member of the genus is the Mangrove Snake, *Boiga dendrophila*. This handsome animal is fairly common in coastal districts and, as its name implies, in mangrove swamps. It grows to about 1.8m (6ft). A generally placid snake, it nonetheless has an unmerited reputation for tetchiness. Elsewhere, other members of the genus are known as Cat Snakes.

Chrysopelea is a genus of very interesting back-fanged colubrids from the Far East whose members are generally called Flying Snakes. Both *Chrysopelea ornata* and *Chrysopelea paradisi* are small, slim arboreal snakes, mainly pale greenish grey in colour overlaid with a net of fine black lines. *Ornata* also has a row of red spots along the spine giving it a most handsome appearance. Like other snakes they cannot really fly, but if they find themselves high up in the branches, needing to escape suddenly, they have an astonishing ability to suck in the underside and spread the ribs out to give themselves a wide, hollow-bottomed appearance. On the cushion of air trapped thus beneath them they are able to glide in a somewhat haphazard fashion for quite considerable distances.

Dryophis is a genus of Asian snakes, pencil slim and attractively elegant, that hunt among the branches of the trees in which they live for small lizards upon which they prey. Their place in the New World is taken by Vine Snakes, *Oxybelis*, whilst other American species of back-fanged snakes are the Lyre Snake, *Trimorphodon*, the Black-headed Snake, *Tantilla*, and the Mussurana, *Clelia clelia*, an interesting species that feeds on other snakes, including pit vipers. Several kinds of snake feed on others, and though none of them is immune to venom from other snakes, it often seems to have far less effect on them than it would on species that are not designed to eat other snakes.

The semi-aquatic Mock Vipers, *Psammodynastes*, live on frogs. Their main claim to fame is that they are one of the snakes that are frequently known as Hoop Snakes in America. The story goes that they are able to take their tails in their mouths at the top of a hill, thereby forming a hoop, and then go merrily bounding down the slope until they reach the bottom or, presumably, hit a rock halfway down. They are also said to have a venomous sting in the tail. There is no truth in any of this, though the latter story must have come about because they do have a hard pointed end to their tails.

Many animals around the world take advantage of the fact that other animals of similar appearance are known by predators to be dangerous. By mimicking the colouration of these dangerous species, harmless ones can often get away without being eaten. One such is the False Coral Snake of South America, *Erythrolamprus*, which mimics the highly venomous true coral snakes. Another fascinating oddity among the back-fanged serpents of South America is the burrowing species *Stenhorhina*.

Though most venomous colubrids are members of the sub-family Boiginae, there are exceptions. The sub-family Homalopsinae contains a variety of aquatic species, of which the most peculiar must surely be the Tentacled Snake, *Erpeton tentaculum*. It is a rough-skinned little beast hardly more than .3m (1ft) in length, with two small tentacles at the front of the head. The Dog-faced Water Snake, *Cerberus rhynchops*, and members of the genera *Fordonia* and *Enhydris* are part of the same group. Both *Cerberus* and *Fordonia* live in crab holes, and indeed the latter eats crabs, which seems an unlikely diet for a snake. But what is perhaps even more interesting is that their venom is particularly effective on crabs, but has comparatively little effect on frogs or small mammals. Members of the Homalopsinae are also odd in that they can swallow prey under water. Another small group of back-fanged colubrids belongs to the Aparallactinae. These snakes of the genera *Atractaspis*, *Aparallactus* and *Xenocalamus* are collectively known as Mole Vipers, despite the fact that they are not related to the true vipers at all, nor are they dangerous.

All the back fangs referred to so far are generally regarded as no sort of threat to man, and indeed if any of them does manage to inject venom the effect is almost always negligible. The Mangrove Snake may be an exception. It is certainly as well to treat it with respect as it is a fairly large snake, but there seems to be no evidence that it has ever bitten anybody. The two real baddies in this group are the Vine Snake, *Thelotornis kirtlandi*, which is sometimes also referred to as the Bird Snake, and the Boomslang, *Dispholidus typus*. There is no doubt at all that either of these snakes, but particularly the Boomslang, can kill a man; but having said that, it should be stated that they are the most mild mannered and placid of snakes. Nevertheless, it is most disconcerting when climbing through the topmost branches of a swaying prickly tree to meet an angry Boomslang coming in the opposite direction at top speed. They are totally arboreal, and come in two colour forms. One is a nondescript brownish-grey, and the other is the most beautiful vivid green. A newly sloughed Boomslang looks as though it's a freshly polished enamelled jewel. Both the Boomslang and the Vine Snake, when threatened, are apt to hiss loudly and inflate their necks to show the skin between the scales - a most daunting and effective display.

The Elapids: Cobras and their Relatives The best known of all the elapids are the cobras, which inhabit most areas of open country from

Pakistan eastwards. They are especially common in agricultural areas where they can take advantage of the rodents that congregate to feed on the crops. They are not at all averse to small birds either, and will take them from nests if they can find them. Cobras are also found in Africa, but even more so throughout Asia. There are numerous varieties of the Common Indian Cobra, *Naja naja*, though they all look pretty much the same – usually blackish-grey in colour, creamy yellow underneath with black bands across the throat. When encountered in the wild, they can look surprisingly like the Indian Rat Snake, *Ptyas mucosus* (a completely harmless species), apart from the cobra's stripes across the throat. When one is snake hunting in the tropics, all too commonly what one sees of an animal is the last few inches of tail disappearing beneath a bush. A common misconception is that all cobras exhibit a typical spectacle mark on the erect hood. In fact one finds everything from this pattern to two separate rings, to a single ring, and just about every variation. Colour variants are sometimes encountered, and a baby albino cobra must be one of the most beautiful snakes in the world.

Cobras live in holes in the ground and although they can climb a short distance into a shrub when they are after a bird's nest, it is unusual to see them anywhere but on the ground. These snakes will put up with a fair amount of abuse before they will strike, doing all they can in the meantime to get away. An angry cobra, nonetheless, is an animal to treat with care. If they are disturbed in their holes they will hiss loudly and deeply in warning, but if in the open when threatened, a cobra will pack its coils neatly around itself, spread its hood and turn to face the enemy. It thus looks alarming, but the range of its strike is limited and in this situation it will often strike with mouth closed in an attempt to dissuade an aggressor from attacking. It should not be thought that a cobra can only bite when the hood is erect. One race of the Indian Cobra, *Naja naja sputatrix*, has developed a spitting mechanism, a technique more frequently encountered in Africa.

According to the literature, the King Cobra, *Ophiophagus hannah*, is to be found in India. If this is so, it must be right on the edge of its range, for although nowhere common it is far more frequently encountered farther east. It is a huge snake, up to 4.9m (16ft) long, commonly a pale brownish-grey colour, and despite a reputation for ferocity it is not generally as aggressive as portrayed, though there are occasions when it can be. King Cobras are frequently encountered in pairs, which no doubt gave rise to the myth that anyone killing a King Cobra will find that its mate chases them to exact revenge. It is much more of a forest species than the previous one, and together with *Pituophis*, the North

American Pine Snake, it appears to be the most intelligent of all the snakes, though intelligence among snakes is only relative. However, it shows far more interest in what is going on than most other species and in captivity it will watch any sort of activity outside its tank. King Cobras are highly specialised feeders, existing only on other species of snakes, which makes them a rarity in zoos outside the tropics.

In Africa the commonest cobras are the Egyptian Cobra, *Naja haje*, and the Black-necked Cobra, *Naja nigricollis*. They are fairly similar in appearance and habits to the Indian Cobra though their hoods, when erect, are not as magnificent as those of the Asian species, as they are narrower. Just as the King Cobra is to be found in forested areas in Asia, Africa too has a Forest Cobra, *Naja melanoleuca*. Perhaps the most interesting cobra of Africa is the Spitting Cobra or Ringhals, *Haemachatus haemachatus*. This handsome brick-red animal has the opening of the venom ducts on the front of the fangs so that when faced with an enemy, the Ringhals, rather than biting, applies muscular pressure on the venom glands to force the liquid in a stream straight at the enemy's eyes. The aim of these snakes is absolutely deadly, and since they have a range of up to 3m (10ft) they can be a real threat to man. The venom causes excruciating pain in the eyes of the victim, and unless flushed out promptly and treated, can even cause blindness. A Spitting Cobra can of course bite, as it must to obtain its prey, but it is very reluctant to do so in defence. One often hears stories in Africa of nests of cobras, but in fact the only cobra to build any sort of a nest is the Asian King Cobra. The female will coil round her eggs to incubate them and, with her head, pull a heap of leaf litter around herself. In due course the nest ends up with what seems to be two chambers, one for the eggs, and above it another in which sits the mother.

Other African species are the Desert Cobra, *Walterinnesia aegyptia*, and an aquatic species known as the Water Cobra, *Boulengerina*, a fish-eating species from central Africa. The notorious mambas are other African examples of the Elapidae. There are generally thought to be only two species, the Black Mamba, *Dendroaspis polylepis*, and the Green Mamba, *Dendroaspis angusticeps*, but there are two others, the Western Green Mamba, *D. viridis*, and Jameson's Mamba, *D. jamesoni*. The mambas are thought to be the fastest snakes on earth, and one of them has been timed for a very short distance at 13km (8 miles) an hour, which might not seem fast, but compared with other snakes is extremely rapid. For

(right) *The Chile Rose and other large tarantulas are found in the USA and through to South America*

years there have been tales about the legendary speed of the mambas, and it was even said quite seriously at one time that an angry Black Mamba could overtake a man on a motorcycle. The reality is something of a disappointment. All the mambas are long slim snakes which are encountered on the ground or in trees and shrubbery. The Black Mamba is not really black, and looks rather similar to the Boomslang. The scales of Jameson's Mamba are outlined in black. All the mambas are aggressive animals, and being fast they are very dangerous.

Perhaps the prettiest of all the elapids are the coral snakes. As with many animals, common names can be confusing and there are coral snakes from various parts of the world; but when most people refer to coral snakes they mean the handsome red, black and yellow animals from the Americas. Though they have been described as aggressive, they usually seem to be the most placid of animals. Apart from the Giant Coral Snake, *Micrurus spixi*, from South America, that grows to 1.5m (5ft), most representatives of the family are rather small. The two commonest, or at any rate best known, species, are the Arizona Coral Snake, *Micruroides eryxanthus*, and the Eastern Coral Snake, *Micrurus fulvius*. These small unassuming animals at one time caused agony and death to the workers in sugarcane fields where they take up residence during the growing season. Although brightly coloured, they are surprisingly difficult to spot and it is easy to understand how the reptiles were trodden upon in the deep leaf litter on the ground when workers wore no shoes. This was a common place to encounter another dangerous snake, the Fer de Lance, *Bothrops atrox*, which is not an elapid but a pit viper. These were better camouflaged beneath the ripe sugarcanes at harvesting. Nowadays when a field is ready to be cut the whole thing is set on fire to burn off all the dead foliage. Tremendous heat is generated for a very short time and great billows of flame explode in all directions, some of the canes adding to the drama when they go off with a bang. It is all very spectacular, and at the end of it the snakes that had been living happily in the plantations are burnt to a cinder.

On the other side of the world, in and around Malaysia, snakes of the genus *Maticora* are also known as coral snakes. *Maticora bivirgata*,

(top left) *The Red Kneed Tarantula looks dangerous but is in fact a most placid animal*

(below left) *This spiderling curiously has pink legs and black feet. When adult the colours are the other way round*

the Blue Malayan Coral Snake, is a truly handsome animal which grows to 1.5m (5ft) in length. The back is dark blue and there is a most vivid blue stripe the length of each side. The head and underside are a brilliant coral red. It is to be found in forested areas where it feeds on lizards and snakes. Though it has only a small mouth, the animal is very venomous.

Some of the best known of the elapids are the kraits. They are chunky, terrestrial snakes that are decidedly triangular in cross-section. During the daytime they are so placid that they can be easily underestimated, but when encountered at night, for they are nocturnal, they are very dangerous. In the odd event of being discovered during the day, they will bury their heads beneath their coils and sit quietly in the hope that the danger will go away. They are so reluctant to do anything else that they will put up with kicks and blows even to the point of death without taking any sort of action.

The Maticora coral snakes, incidentally, have enormous venom glands, far bigger for their size than any other group of snakes. The glands extend back into the body for a third of its length, taking up so much space that the heart is displaced backwards to make room for them. Despite this, deaths from Maticora seem to be unrecorded. There is apparently only a single authenticated case of someone being bitten, and that was a Dr Jacobsen of Java who wrote that a single fang from a Maticora had penetrated the skin between the index and middle finger of his hand. At the time of the bite he only felt a mild local pain, which lasted for about two hours. Then he suddenly suffered an attack of dizziness, and found that he had considerable difficulty in breathing. There were five or six of these attacks, each lasting five to ten minutes, and afterwards he recovered. He later speculated that had he received a full dose of venom from both fangs the picture might have been very different.

Biologically, Australia is an odd place. It is full of all sorts of species of animal that occur nowhere else, and from a herpetological point of view it is every bit as strange. There are more venomous species of snake in Australia than non-venomous ones, and most of them are elapids; vipers and pit vipers are completely absent. Though the Australian species of venomous snake generates much superstition and panic, many of the elapids from that part of the world are relatively harmless, their bites causing no more damage than the sting of a wasp. Having said that, some Australian snakes are every bit as dangerous as snakes in other parts of the world. Of the eighty-five species of elapids in the country, the Taipan, Oxyuranua scuttelatus, is by far the largest, reaching a length of 4m (13ft). A bite from one of these snakes can kill a man in minutes. Luckily the number of casualties is not too high since the species is rare and found only in comparatively undisturbed parts of north-eastern Australia. There is a closely related species, the Inland Taipan, *Oxyuranus microlepidotus*. Two other really dangerous snakes in Australia are the Death Adder, *Acanthophis antarcticus*, and its close relative the Desert Death Adder, *Acanthophis pyrrhus*. Mortality from the bites of these two species is remarkably high, around 50 per cent. The two Tiger Snakes, *Notechis scutatus* and *Notechis ater*, give cause for concern in large parts of Australia, and the six species of Brown Snake. *Pseudonaja*, are also highly poisonous and killed whenever they are found. The Tiger Snakes are notable in giving birth to live young and are remarkably prolific, producing up to seventy-two babies at a time.

The Hydrophidae: Sea Snakes At one time the sea snakes used to be lumped with the elapids; nowadays they have a family to themselves. They are to be found in tropical seas eastwards from the eastern coast of Africa as far as the western coast of the Americas. Strangely, they do not occur at all in the Atlantic. The greatest concentrations exist around the northern coasts of Australia, the Philippines and Indonesia where they can sometimes be found gathering in enormous quantities. Great rafts of many thousands of sea snakes may sometimes be encountered at sea. It has been suggested that this might be because small fish frequently congregate beneath floating debris.

Sea snakes have been killed for their leather for a long time, but only in small numbers that left the populations unharmed. The situation may be changing. When the film *Crocodile Dundee* became popular a few years ago, the star, Paul Hogan, was shown wearing a jacket made from the skins of sea snakes. Instantly the world demand for this product jumped alarmingly, and continued to rise so that today there is some concern about the continued existence of sea snakes.

There are about fifty species, all well adapted for their marine existence. The tails are flattened from side to side as an oar, and in some species the whole body tends to be laterally flat. The largest species is *Laticauda semifasciata*, which gathers in thousands at the islands of Gato, north of Cebu, in order to breed in the large coastal caves that are open to the sea. All sea snakes produce live young whilst at sea. Some species such as the Amphibious Sea Snake, *Laticauda colubrina*, do not have an entirely marine lifestyle as most species do. They live along the coast and enter the water frequently. Like the others, however, they are well developed for an aquatic life. The eyes and nostrils of sea snakes are set at the top of their heads and some of them, like the Banded

Slender-necked Sea Snake, *Hydrophis fasciatus*, and the Small-headed Slender Sea Snake, *Microcephalophis gracilis*, have one of those elegant design features that one sometimes comes across in nature, which renew one's appreciation of how clever the world of biology is. These snakes are almost plesiosaur-like in shape, in that they have very slender necks, but further down the body is much wider and chunkier. The girth of the abdomen of *Microcephalophis* for example is four to five times that of the neck. This enables the animal to strike far more efficiently than otherwise it might. Being underwater when it does so, there is nothing for it to hold onto to obtain purchase, so the large heavy lump at the back end provides a fairly inert base from which to strike.

Sea snakes seem to be inoffensive creatures which will put up with considerable abuse before biting. Colonel Wall of India said that there were scarcely any records of casualties even among the fishermen who haul them in by the dozen every day when they become tangled in the nets. The sea snakes so caught were handled fearlessly by these men and when they disentangled the snakes they simply threw them back in the sea. Their placid nature is disputed by some, but most people accept that they pose no great threat, which is as well since sea-snake venom is extremely toxic to man and, despite Colonel Wall's fishermen, they should not be treated lightly. The venom is very fast working and even a large eel stiffens and dies in only a few moments. This rapidity of action is important to sea snakes as it does not permit a bitten fish to disappear into an inaccessible crack in the coral before it dies.

The Viperidae: Vipers and Pit Vipers The vipers are regarded as the most modern group of snakes. There are about 150 species altogether divided into two subfamilies. The first of these is the subfamily Viperinae, which includes snakes like the common European Adder and the notorious African Puff Adder. Snakes from this group are to be found throughout Europe, Africa and Asia. The other subfamily is the Crotalinae or pit vipers. They occur in greatest numbers in the Americas, though they are also found in south-eastern Asia. The best known members of this group are without doubt the rattlesnakes. Most snakes in this family give birth to live young and most species are ground- dwelling though some from Asia are arboreal. They all have a typical viper shape: a short, stubby body, a narrow neck and an arrow-shaped head. They have small scales, often keeled, giving a rough feel to the snake in contrast to most colubrids and elapids which are smooth and glossy.

The European Viper (or Adder), *Vipera berus*, occurs farther north than any other species of snake, even into Norway in fact where it manages to survive by hibernating through a very long winter. The most northerly of the Adders tend to be darker than many of their more southern relatives as this encourages the absorption of heat during the short summer months. Melanistic Adders can be found elsewhere, but most of them come from the northern part of the range. At the other end of the globe another viperid, the Snout-headed Lance Head, *Bothrops ammodytoides*, is the most southerly snake of all, occurring well down into Argentina. The smallest viper of all is perhaps Orsini's Viper, *Vipera ursini*, of southern Europe, which reaches a length of 30cm (less than 1ft). The Bushmaster, *Lachesis muta*, on the other hand is a whacking great snake of 3.6m (12ft) or more from South America.

The European Viper is the only venomous snake in the British Isles, where it is regarded with much fear and loathing. Yet it is a most inoffensive little animal, far preferring to disappear silently into the undergrowth than to bite. Despite people's attitude towards it, hardly anyone has actually died from Adder bites in Britain, and of those that have most have been small children or adults with heart conditions. Even though the Adder is the commonest reptile in Britain, most people are not lucky enough to see one of these attractive little animals; as a rule one has to be an experienced herpetologist. The best time is early morning in spring. Find a good Adder area and arrive soon after the sun has risen. When you set out from home in the cold darkness you will feel that the last thing you want to see is a wretched snake, but once you have been up half an hour with a cup of coffee inside you and can watch the early sun catching clumps of primroses from low on one side and glinting off water in the car tracks, you will suddenly realise that despite the government of the day, Britain is a super place to be. That is the time to go for a quiet wander round and if you are lucky you will see a few snakes creeping out to enjoy the warmth of the early sun. They will usually find themselves a small exposed place like the top of a hummock of sand where they will make themselves comfortable for a while. As the sun soaks into their little bodies you can watch them spread themselves flat so that as much surface area as possible is exposed to the warmth.

The first to emerge each spring are the males, followed some days later by females. Mating takes place at this time and sometimes it is possible to see two males engage in a ritual fight. They twist the front part of their bodies together in a vertical heaving column of snake, each trying to push the other over. It is this behaviour that gave rise to the two entwined snakes on the staff of Hermes the Messenger. It is often confused with the staff of Asclepios, the god of medicine, that also features a serpent, but in that case the snake is the Aescalupian Snake, a non-venomous species.

Many small vipers belonging to the genera *Vipera* and *Cerastes* are to be found throughout Europe and the Middle East. For the most part they are all fairly similar in appearance, though some of them have what appear to be horns on their face, such as the Horned Viper, *Cerastes cornutus*, or *Cerastes cerastes*, which has a horn above each eye that gives it a more exotic appearance than the run-of-the-mill vipers, so much so that snake charmers often use it in their displays. Sometimes when these snakes are not available, a charmer will push hedgehog spines up through the mouth of a hornless species to stimulate the appearance of a Horned Viper. The False Cerastes, *Pseudocerastes fieldii*, was only named in the year 1930, and is said to be the Horned Viper of the book of Genesis.

The Asp, *Vipera aspis*, famous throughout the world as allegedly being the snake that Cleopatra used to commit suicide, has perhaps achieved fame by mistake as there is some evidence that in fact she used an Egyptian Cobra. The Palestinian Viper, *Vipera palestinae*, is a considerable threat in the Middle East, and is responsible for more cases of snakebite in Palestine than any other venomous snake. Many of the islands of the eastern Mediterranean have good populations of snakes, but due to the increase in tourism over the last few years habitats are disappearing fast, and with them so are the snakes. Sadly, little is done in most places to conserve the remaining populations, and in some areas the snakes are deliberately killed whenever they are seen. For many years the Blunt-nosed Viper, *Vipera lebetina*, on the island of Milos was systematically exterminated due to a practice of paying bounties for dead specimens.

There is a little viper found in India and north Africa called variously the Saw-scaled Viper or the Carpet Viper, *Echis carinatus*; it is especially common in north-western India. It is small and well disguised, and consequently causes a lot of damage in an area where hardly anyone wears shoes. Even a newly born snake of this species can kill a man, and these animals will bite without much provocation. Some years ago bounties were paid on this species as well, and over a period of six years about 200,000 snakes were brought in. This depredation does not seem to have made much difference to the population of the area since the snake is still astonishingly common.

When disturbed, the Saw-scaled Viper has a habit of moving its body into an S-shape and rubbing its rough scales together to produce a sound reminiscent of a rattlesnake. The resulting name of 'Sidewinder' is also applied to a number of other species of snake and refers to their rather peculiar method of locomotion. When most snakes move in a given direction they keep their bodies facing more or less the way they want to go. Sidewinders on the other hand travel with a sideways motion, keeping their bodies at an angle of about 45 degrees to the direction in which they are travelling.

Further south in Africa the *Viperas* and the *Cerastes* are replaced to a great extent by the genus *Bitis*. The best known members of this genus are without doubt the Puff Adder, *Bitis arietans*, and the Gaboon Viper, *Bitis gabonica*. The Puff Adder is to be found over much of Africa south of the Sahara, and in Morocco and Arabia wherever there is open country. It grows to a length of about .9m (3ft) and is well camouflaged in a mixture of fawns and browns. It is not a snake to take any liberties with as it will strike without warning and is responsible for numerous cases of snakebite each year. The Gaboon Viper on the other hand wants nothing more than a quiet life. It does not enjoy nearly as wide a distribution as the previous snake, and occurs in forested parts of the continent. Like the Puff Adder it is well camouflaged, but when examined closely one can see that it has far more colours and is more brightly hued than the former. Be in no doubt that the snake can deliver a bite capable of killing a man. It has the longest fangs of any snake in the world, almost 5cm (2in) long, and packs enough venom to kill 200 people. Yet it is so placid that small boys will sometimes pick it up by the tail and carry it, or if the boy is small and the snake is large, drag it along the ground for delivery to animal collectors in exchange for a few coppers. Strangely, these same small boys are terrified of harmless house geckos that they are convinced are deadly. Without doubt the most colourful of the African vipers is the Rhinoceros Viper, *Bitis cornuta*. For a viper it is extraordinary. When freshly sloughed it is exquisitely coated in a pattern of blue and yellow and purple along the back.

An equivalent snake, if you like, to the African Puff Adder is the Russell's Viper, *Vipera russelli*, of Asia, where it is sometimes known as the Tic Polonga or Daboia. It is an attractive, russet-brown snake with a row of dark rings down the back. Aggressive and dangerous, it is normally found only on the ground but it can perform the most amazing feats on occasion, and has been known to climb the small fence surrounding a verandah to remove a happily singing Red-whiskered Bulbul from a bird cage hanging from the roof. With its attractive markings one would have thought that the snake would be in demand for its skin, but this is not the case. Dealers in reptile skins say that the leather of this species is not as tough as that of the Indian Cobra, and consequently there is not much demand for it. 'Not much demand', like everything else when you are talking about this subject, is relative, and in 1983 the United Kingdom applied to import 80,000 skins of this species from India. The range of this reptile extends further east than India, through to that of the pit vipers, the relatives of the rattlesnakes in the New World. As we have seen, they are known as pit vipers because on either side of their face they have small pits with heat-sensitive membranes at the bottom which enable them to locate warm-blooded prey more efficiently than by just using the other senses, and more especially at night when sight is not much use.

The two genera of pit viper in this part of the world are *Trimeresurus* and *Agkistrodon*. Snakes of the genus *Trimeresurus* are not to be messed about with, yet many people in the East treat them in a remarkably cavalier fashion. In fact, on the price list of the biggest dealer in snakes in Thailand, the various species are listed under headings relative to their venomosity. Cobras and the like are listed as 'Highly Poisonous', but *Trimeresurus* comes under the heading 'Mildly Poisonous'. Of course they vary, and the common Wagler's Pit Viper, *Trimeresurus wagleri*, is kept in numbers at a temple in Penang in Malaysia where they are handled regularly by many people. Occasionally one hears of a tourist being bitten, but by and large man and snake seem to get on well together. The most dangerous pit viper in Asia is the Habu, *Trimeresurus flavoviridis*, of Japan. It grows to a length of 1.5m (5ft) and was greatly feared by the American forces in the Far East during World War II. Another, rather smaller relative of the Habu is the Kufah, *Trimeresurus okinavesis*, which is also dangerous. Most of the Asian species of *Trimeresurus* are arboreal, the terrestial niche being occupied by the other genus of this group, *Agkistrodon*. *Agkistrodon rhodostoma* is known locally as the Malayan Moccasin. It lays 12-30 eggs in a nest and guards these until they hatch. Another *Agkistrodon* that has a large range throughout the continent is *Agkistrodon halys*, which is to be found from south-east Russia as far as Japan. Asia is a great place for venomous snakes. They are varied in form and colour and habits, and in some instances still common.

The genus *Agkistrodon* is represented on the other side of the world in North America by two species, the Copperhead, *Agkistrodon contortix*, and the Cottonmouth or Water Moccasin, *Agkistrodon piscivoros*. Of the 19 venomous snakes in the US, 17 are members of the pit viper sub-family. The Copperhead is widely distributed through central and southern areas of the United States in wooded hillsides or in areas near water. It can be seen basking during the hours of daylight in spring and autumn, but during the warm summer months it becomes nocturnal. Copperheads feed on almost any animal of suitable size that they encounter, and though a bite from one of these snakes is extremely painful, it is unlikely to result in the death of an adult human. The Cottonmouth

Moccasins are instantly recognisable from their appearance. Originally they were made by the north American Indians from the skins of Water Moccasins (John Nichol)

is found throughout the south-eastern part of the US. The species is never far from water, in which it spends a fair amount of time hunting the aquatic animals on which it feeds. It swims with its head well above the surface, and although one might come across it basking in the sun, the snake is far more active at night. It is infinitely more dangerous than the preceding species and should not be approached. When it is threatened the Cottonmouth, unlike most other snakes, will stay put and face you, opening wide its mouth to reveal the almost white lining which gives it its common name.

The remainder of the North American pit vipers are all rattlesnakes. To look at, rattlesnakes are much like any other viper except that they almost all have a rattle at the end of the tail formed by a series of dry, horny segments that vibrate with a buzzing or rattling sound when shaken by the animal as a warning of its presence to a predator. Every time the snake sloughs, a new segment is added. The largest of all the rattlesnakes is the Eastern Diamondback Rattlesnake, *Crotalus adamanteus*, which can grow to a length of 2.4m (8ft). Together with the Western Diamond

Rattlesnake, *Crotalus atrox*, it is often regarded as the most dangerous snake in the United States. Numbers of the Eastern Diamondback, which lives in Florida and the nearby states, have been depleted over the years by the loss of suitable habitat as building work has stretched across the area, and by rattlesnake hunters taking their toll. *Crotalus cerastes* is known in North America as the Sidewinder, for the same reason that *Echis carinatus* is called by that name in India, in other words, because of its curious and highly specialised method of locomotion. The 1.2m (4ft) long Mojave Rattlesnake, *Crotalus scuttatus*, has become well known to herpetologists as the rattlesnake that causes more respiratory distress in humans who have been bitten than any other species. The Western Rattlesnake, *Crotalus viridis*, is an aggressive, irascible animal. In the northern part of its range large numbers are sometimes found in hibernacula awaiting the onset of warmer weather.

There are two species of pigmy rattlesnake in the United States, but although they are referred to as pigmy, in fact they are not particularly small, growing to between .75m and 1.2m (2.5 and 4ft). The Massasaugu, *Sistrurus catenatus*, was so named by the Chippewa Indians. The name means Great River Mouth and probably refers to the fact that this snake is often found in the low-lying marshland that surrounds the mouths of rivers. The Pigmy Rattlesnake, *Sistrurus miliarius*, is sometimes known as the Ground Rattler. Some authorities say that it is very short tempered and will strike without provocation while others claim that it is extremely placid. A well known American entomologist, W.T. Davies of Staten Island, was bitten by the latter species and described what happened.

Collecting insects one day he was bitten by what he thought was a harmless juvenile Hog-nosed Snake. He found that there were two small punctures near the nail of the third finger of his left hand. He caught the snake, which was later found to be a Pigmy Rattlesnake. He immediately felt considerable pain at the site of the bite, which leaked blood but, believing as he did that the attacker was of a harmless species, he continued collecting. Four hours later his arm felt heavy and sensitive, and was swollen badly. Worried, he called a doctor who agreed that the snake must have been fairly harmless and that there was nothing to worry about, but just to be on the safe side he applied some iodine to the tiny wounds and went home. Davies's arm was painful throughout the night, so much so that he could not sleep. Eventually as it became light he got out of bed, and promptly fainted. When he recovered consciousness he discovered that the bitten finger had turned black in the night and the wounds were leaking blood and serum. He fainted again. The pain went on for what must have seemed an everlasting forty-eight hours, by which time his whole left side was swollen and the underside of the arm was black and blue, swollen and flabby. By the third day the swelling had begun to go down, and the pain was certainly less severe than it had been. By the fourth day he was feeling infinitely better and by the eighth day all the swelling had disappeared, though the finger continued to be useless for the next eighteen months. Don't get bitten by a Pigmy Rattlesnake!

One odd species of rattlesnake known as the Rattleless Rattlesnake, *Crotalus catalinensis*, from the island of Santa Catalina off the Baja California, has only a vestigial rattle. And though the genus *Crotalus* extends down to South America, the largest genus of pit vipers throughout central and South America, and in the West Indies, is *Bothrops*. Many snakes of this genus are called colloquially Fer de Lance, but the true Fer de Lance, *Bothrops lanceolatus*, is only to be found on the French West Indian island of Martinique. Similarly in South America several completely different snakes are known as Jararacussu or Jararaca, but the true Jararacussu is *Bothros jararacucu*, and the real Jararaca is *Bothrops jararaca*. The pit vipers of this region have colonised all sorts of habitats, and changed slightly to fit their environments, so for example, there is a Prehensile-tailed Pit Viper, *Bothrops schegeli*, and there is even a very fierce Central American species known as the Tamagasse or Jumping Viper, *Bothrops nummifer*. It really can jump, too. It throws itself forward so violently that it actually clears the ground. The largest of the pit vipers from this region is the Bushmaster, *Lachesis muta*. It can grow to over 3.6m (12ft) and when it is ready to strike it looks a really formidable opponent. Somehow one is not surprised at thin snakes being long, but to find a long viper comes as a shock. There is an awful lot of a big Bushmaster. The prize for the snake with the most appalling name must go to the Wutu, *Bothrops alternatus*!

LIZARDS

There are approximately 3,000 species of lizard in the world, but would you believe that only two of those are venomous? That really seems a biological oddity when one considers that the vast majority of lizards are carnivores. Yet it is true that only two have developed this special feature, and both of them are members of the same genus *Heloderma*. Indeed they are the only two members of that genus, and they occupy a small area in Utah, Nevada and Arizona, and south into the adjacent parts of Mexico. Within this range one species, the Gila Monster, *Heloderma suspectum*, is confined to the United States, and the other, the Beaded Lizard, *Heldoderma horridum*, lives in Mexico. Both animals are short, stubby lizards about 49 to 74cm (19 to 29in) long, covered in small knobbly scales. Both are a mixture of black, yellow and pink. Unlike many species, the heloderms are not fast active lizards, preferring to remain underground in a hole that was previously occupied by something else, or in one that they have dug themselves. They emerge to hunt for food, and being slow moving they prefer food that cannot run away such as the chicks of ground-living birds and their eggs. It seems that they locate their prey mainly by using a sense of smell rather than sight, and they can be observed in the wild following what appears to be a scent trail with their heads down in the manner of a dog.

An experiment was made on one occasion to try and discover how relatively important sight and scent were in the location of food. An egg was placed in front of an animal until it had recognised its presence, and then the egg was moved over the ground along a twisted track, and placed a metre (yard) away. The lizard appeared not to be able to see the egg at all, but it soon became clear that it was not going to have any trouble tracking it by scent. It carefully followed the trail to the bitter end even though at one stage it was walking along only 13cm (5in) away from the egg.

Both Gila Monsters and Beaded Lizards mate during summer and lay between three and five eggs in autumn and winter. These delightful animals are mainly nocturnal though one can also observe them on a warm day, especially if it is damp. In the deserts where they live, they make full use of any rain that falls, and eat whenever they can, storing surplus food in the form of fat in the tail.

The venom is not injected as with most snakes, but when a heloderm bites, which incidentally it can do at the speed of lightning, it grabs the food (or the predator) and will not let go. While it is doing this it chews, for the teeth are grooved to allow the venom to flow from the glands at the base of the mouth to the damaged tissues of the unfortunate victim. A heloderm will hang on with the tenacity of a bulldog and is the very devil to dislodge once it has taken hold. Usually the venom is not fatal to human beings though the bite is very painful. Only eight cases of death have been recorded as due to these powerful lizards, and one is that of a fairground barker who had a history of drug addiction and heart trouble. One day when he had consumed a considerable quantity of whisky he wanted to show a friend that Gila Monsters were not as dangerous as they had been portrayed. To make his point he prised open the mouth of a captive specimen, inserted a finger and waited to see what would happen. The Gila Monster was delighted to have been chosen to take part in the demonstration and chewed away with gusto. The demonstrator died.

AMPHIBIANS

FROGS, NEWTS AND SALAMANDERS

It would seem that most amphibians secrete some toxic substances in their subcutaneous tissues. None of them is able to inject venom into humans; when roughly handled, however, plenty of them can release these compounds through ducts in the skin. Most of them are so mild that a human who handles one of these animals is not even aware that the liquid is toxic, though without doubt it does deter predators from eating the amphibian. A few amphibians secrete toxins powerful enough to cause a reaction if a person's skin is sensitive to such things, or to cause discomfort if the stuff gets into an eye or mouth. Even when this happens the effect is hardly more than a degree of discomfort. Several animals might fit into this category. The Pickerel Frog of North America, *Rana palustris*, is one such, though the secretion is effective enough to kill any other frogs kept in the same tank. Ridiculous though it sounds, there is even a venomous newt. The Californian Newt, *Taricha torosa*, secretes a venomous substance through its skin that has been known to cause a local reaction in humans. When disturbed this newt like many others from the same part of the country throws its tail up to reveal the bright orange underside to deter predators, rather like the beautiful little Yellow-bellied and Fire-bellied Toads of the Old World do to dissuade any attempts at interference. Another species that is said to affect some people is the European Fire Salamander, a large black and gold amphibian with a wide distribution.

There are also amphibians that secrete skin secretions so toxic they can easily cause human death, and some of these substances are among the most virulent known to man. The animals responsible are inoffensive beasts, perfectly content to live out their lives in the rainforests of South America. Long ago the Amerindians discovered that if one takes an arrow-poison frog and sticks a spine through him and holds him over a fire, a liquid starts to ooze from his skin. When this is scraped off and left to ferment, it becomes a perfect substance to coat the tips of arrows so that they are lethal, whatever part of a target animal they hit. Such an opportunity was not to be missed, and since then various tribes in the northern part of South America have tipped their arrows in this way before going hunting or to war. In other parts of South America the juices from various plants are employed to produce the same effect, the best known being curare. A monkey that is shot for the pot using an arrow tipped with poison from a frog is perfectly harmless, and can be eaten with no ill effects.

All the arrow-poison frogs are beautifully coloured, and very small. They can usually be handled carefully without harm, and provided they are not excited there should be nothing to worry about, but as with all such things they need to be treated with care as they can be very dangerous. Although there are quite a number of species, it seems that only about half a dozen are actually used for the treatment of arrows. The preparation of arrows in this fashion still continues though with the availability of firearms it is dying out, for which the poor old frogs must be grateful since it takes the venom from fifty of them to tip a single quiver of arrows. The species most commonly used in this way are:

Flat-spined Atelopus, *Atelopus planispina*. It is gold and green with white spots and black markings, and comes from the mountains of Ecuador.
Zetek's Frog, *Atelopus zeteki*. Gold all over, the male has black blotches. It comes from Panama.
Yellow-spotted Arrow-poison Frog, *Dendrobates flavopictus*, is a black species with bright yellow spots and lines, from the uplands of Central Brazil.
Boulenger's Arrow-poison Frog, *Atelopus boulengeri,* is black with cream spots and comes from high in the mountains of Peru and Ecuador.
Three-striped Arrow-poison Frog, *Dendrobates trivittatus*, a black and yellow striped species from the northern part of South America.
Two-toned Arrow-Poison Frog, *Phyllobates bicolor*, mainly a vivid red with black markings which also comes from a wide area of northern South America.
Gold Arrow-poison Frog, *Dendrobates auratus*, a delightful little gold and green frog from Nicaragua through Panama to Colombia.

All the arrow-poison frogs inhabit the rainforest and can be found wherever there is sufficent water in which to lay their eggs. They are such tiny animals that the smallest volume of water will suffice and it is not unusual to come across them living in the central vase of epiphytic bromeliads high above the forest floor. To survive in such a competitive, over-populated environment as a rainforest one has to take advantage of everything one can, and some aspects of the animals that live there have become highly specialised. The male Two-toned Arrow-poison Frog, for example, carries his tadpoles around on his back where they hang by their mouths until they are large and developed enough to lead independent lives. To prevent them drying out, the father periodically immerses himself in water.

TOADS

All toads of the genus *Bufo* secrete venom in the parotid glands behind the eyes. In daily life there is no sign of it, but in a stressful situation a toad will produce a white froth, which like all animal venom is a complicated chemical cocktail. That produced by toads contains compounds known as bufonin and bufogin as well as bufotalin. They work in the same way as digitalis in that they increase the tonicity of the heart to the point of stopping it, and cause weakened breathing, muscular paralysis and nausea, though all these symptoms usually only happen in small animals. Most toads are not harmful to man, nor even to princesses who kiss them. However, there are always exceptions. In 1987 it was reported that a Brazilian naturalist had died as a result of toad venom, and it is known that the hands of people who have handled the West Indian tree frog, *Hyla vasta*, and the African frog, *Phrynomerus bifasciatus*, have occasionally developed a rash.

FISH

Not many freshwater fish are venomous, but in the seas around the world there are varieties of fish which can cause much pain and even death to man. Probably the fish that is the greatest threat is the Weever, or rather, the Weevers, as there are several different animals in this group. The fish known by that name in Britain is *Trachina vipera*. This particular Weever is to be found around the British Isles and Europe as far south as Morocco. It lives in water up to 50m (160ft) deep though it is most commonly found in less than 6m (20ft) where there is a bottom of clean sand. As a result it is not surprising that each year there are many reports of swimmers being injured by Weevers. The fish lies buried just below the surface of the sand, and anybody stepping on it will soon regret it as venom is injected into the foot by the spines on the gill covers and the first dorsal fin. Stings from Weevers are extremely painful, and although they are unlikely to result in death, medical assistance should always be sought. It is said that a useful first-aid measure is to apply urine to the site of the stings. Weevers hunt for their food at night, catching small crustaceans and fish to eat, during which operation they make no use at all of their venom spines which are purely for defence.

The Mediterranean is particularly rich in weevers of various species, such as the Spotted Weever, *Trachinus araneus*, and the Greater Weever, *Trachinus draco*. This latter fish is frequently sold as food in Europe, especially in Belgium and the Mediterranean countries, after the spines have been removed by the fisherman who catches the animal.

There are several species of venomous fish that would probably not cause any harm were it not for the fact that they are extremely beautiful and flamboyant, and therefore are considered to be desirable aquarium species. They are variously known as Lionfish, *Pterois volitans*, or Zebrafish, *Dendrochirus zebra*, or sometimes Dragonfish, Red Firefish, Tandan, Butterfly Cod, Rock Perch, Lolong or Turkey Fish, which is why it is a good idea to stick to scientific names! Whichever name you care to use, these fish have eighteen venomous dorsal spines. Stings from these fish, as with weevers, are very painful and can cause convulsions, delirium and heart failure, though death is rare. The spines sometimes break off in the wound and this can lead to secondary infections and sometimes gangrene. They are not aggressive fish, however, and are perfectly content to drift slowly and elegantly round the coral reefs in which they live. These delightful little animals are sometimes known also as Scorpion Fish, but the true Scorpion Fish, such as *Scorpaenopsis diabolus*, is completely different. It is well camouflaged and can sit among the rocks and marine growths where it lives, completely invisible, but it has flash colours on the inside of the pectoral fins and when it flashes these vivid reds and yellows the effect is quite startling. The spines on the head and the dorsal fins are venomous. Scorpion Fish are widespread throughout the Indo-Pacific region and are found especially in Hawaii.

Several fish of the family Siganidae, commonly called Rabbitfish, which also occur throughout the waters of the Indo-Pacific region, are venomous. Most of them feed on algae on the reefs where they live, and frequently come into very shallow water. In some parts of their range they are caught for food but in other places they are regarded with repugnance. The Rabbitfish have strong, sharp spines on their fins which are responsible for painful stings. In the Far East some Rabbitfish, *Siganus javus* and *Siganus canalicutatus*, are known as the Chinese New Year Fish as large numbers of gravid females are caught in February, the time of that festival. One Rabbitfish, *Siganus rivulatus*, is interesting because it has swum up the Suez Canal from the Red Sea and become established in the eastern Mediterranean.

Stonefish, *Synanceia horrida* and the closely related Devilfish, *Synanceia verrucosa*, have erectile venomous spines in the dorsal fins, which cause much pain and distress to man, even death. Though they are often found on and around coral reefs, they also live in estuaries and mud flats.

Many of the rays belonging to the families *Dasyatididae* and *Myliobatididae* are venomous. The former are commonly known as stingrays,

Puffer fish awaiting preparation in a Fugu restaurant (Herbert Lieske)

and the latter as eagle-rays. Rays of these types have long, whip-like tails which are equipped with one or more venomous spines near the base. Normally placid, most accidents with these animals happen when a victim treads on them while they are lying camouflaged on the bottom. The disturbed fish will thrash the tail about, and the result is often a very nasty sting. There are records of these fish using their powerful tails so strongly that the venom spines have penetrated the planking of small boats. Venom is not used at all in food gathering, and the normal diet is more or less any living animal of suitable size. In some areas they cause considerable damage to commercial mollusc fisheries. The Common Stingray, *Dasyatis pastinaca*, is found from Scotland to the Mediterranean. The Southern Stingray, *Dasyatis americana*, is common throughout its range along the eastern seaboard of America from New Jersey in the north to Rio de Janeiro. The biggest stingray of all is *Dasyatis brevicaudata*, which can grow to a length of 4.3m (14ft); an animal of this size can weigh 340kg (750lb). This species has been known to kill human beings and is regarded as very dangerous by fishermen. Some species of stingray such as the Blue-spotted Stingray, *Dasyatis kuhlii*, and Nieuhof's Eagle-ray, *Aetomylaeus nichofii*, are eaten in the Orient in small numbers by the families of fishermen who find them in their nets.

Freshwater catfish of the genus *Pangasius* from Thailand and adjacent countries are capable of inflicting nasty wounds with their venomous spines. *Pangasius sanitwongsae* grows to a colossal 3m (10ft) and is captured throughout Thailand for food, though due to the revolting carrion upon which it feeds it is refused by many. This species has caused the death of fishermen, while the Pungas Catfish, *Pangasius pangasius*, is capable of causing injury though not apparently death. It is frequently kept as an aquarium fish.

It is interesting to note how many common and scientific names of venomous animals indicate how they are regarded by man. Perhaps the most indicative of this attitude is the common name of the fish *Inimicus didactylus* - the Bearded Ghoul.

ARACHNIDS

SPIDERS

It may surprise some people to discover the presence of venomous fish, but it can be no surprise to anybody to read that scorpions and spiders, both arachnids, are capable of injecting venom into man. It seems that more people kill spiders in their homes than any other animal, because they regard them as a pest, and many are frightened of them. It is true that every single species of spider is venomous, but the vast majority are unable to hurt anybody. Certainly no British spider is able to do any harm. Throughout the world there are quite a number that are large enough and venomous enough to bite a human being, but most of these cause no more discomfort than a sting from a wasp or a bee. This is not to say that it is any fun being bitten by them, but they are not as dangerous as is often made out. However, some species of spider are very dangerous and have caused human deaths.

The most notorious spider is probably the Black Widow, *Latrodectus mactans*, which lives in the United States and other parts of North and South America, southern Europe, Africa, Asia, Australia and New Zealand in a whole host of sub-species and geographical varieties. It is only the larger females that cause any problem as the males are too small to be able to penetrate human skin with their fangs. One of the reasons that Black Widows are generally regarded as so dangerous is their habit of taking up residence in and around human habitation. Consequently they are to be found in outside toilets, lean-to buildings, wood piles and so on. These spiders build a hammock of silk from which they hang upside down underneath. Away from man they are said to live amongst rocks in long grass though it would seem that they are likely to be found in a variety of other habitats.

As the animal has such a huge geographical range it is hardly surprising that it has a seemingly endless variety of names. In France it is known as Malmignatte, and throughout Europe its names tend to be similar. In Arabia it is called the Jockey or Red-backed Spider. The New Zealanders know it as Katipo or Night Stinger, and in South Africa it is Swart Knopiespinnekop. Toxicologists and spidermen generally tend to play down the toxicity of the venom nowadays, but according to the literature Black Widows have been responsible for a number of deaths over the years. It is said that there were swarms of the things in Spain in 1833 and 1841, and similarly in Sardinia in 1833 and 1839 when many people were bitten and some died. There is also a case where hundreds of these spiders invaded the town of Raetahi in South Island, New Zealand, after a minor earthquake caused many small fissures to open up in the area. At this time a seven-year-old girl and her brother were bitten and died as a result. Another swarm of Black Widows plagued Yugoslavia every year between 1938 and 1958, and this phenomenon was extensively studied in 1946. During that year, though many people were bitten, there were only two deaths.

The Gosiute Indians of Utah used the venom of the Black Widow, mixed with that of rattlesnakes, to tip their arrowheads.

Another notorious spider is the Funnel Web Spider from Australia, *Atrax robustus*. There are three species of the Funnel Web, but it is this one, the species to be found in northern Sydney, which causes most concern. It first hit the headlines in 1927 when it killed a boy in ninety minutes. Since then there have been numerous reports of casualties, and in 1971 a three-year-old girl was saved by doctors who treated her with atropine after she had been bitten twice.

For a long time it was thought that the Funnel Web Spider was probably the most venomous spider in the world, or at any rate in Australia. Though it will be of no consolation to anyone bitten by this animal, in the toxicity stakes it has been left well behind nowadays. It would seem that the thirteen so-far named members of the genus *Hadronynche* are more of a threat, and the Mouse Spider, *Missulena insignis*, which had been thought to be only mildly venomous, is now acknowledged to be really dangerous; certainly it is far more toxic than the Funnel Web. Two other spiders, the White-tipped Spider, *Lampona cylindrata*, and one of the genus *Lycosa* are currently under investigation in relation to an appalling condition that bitten humans suffer from, known as necrotising arachnidism. Affected tissue refuses to heal for a long time where it has been dissolved by the venom, and when it finally does the victim is left with severe, disfiguring lesions.

Australia is not the only part of the world with dangerous spiders. South America is pretty well provided, and it is sometimes said that the Brazilian Huntsman, *Phoneutria fera*, is the biggest threat to man from the spiders of that part of the world. There are many who would dispute this, however, and a related species known locally as Aranha Armedeira, *Phoneutria nigriventer*, is the largest spider of South America. It hides in clothes and shoes and is quite extraordinarily aggressive. It also has the largest venom glands of any spider, 10.2mm x 2.5mm (0.4 x 0.1in). It bites furiously and determinedly and the venom is so potent that only 0.006mg (0.00000021 oz) is sufficient to kill a mouse. A single specimen has been shown to possess 1.35mg (0.000044oz) in its glands.

Two other South American spiders, the Podadora or Bola Spider, *Glyptocranium gasteracanthoides*, of Argentina, and the Black Tarantula of Panama, *Sericopelma communis*, have been responsible for human deaths. There is a species from the same part of the world that is also known as the Funnel Web Spider, *Trechona venosa*, and although it is potentially capable of killing a man it is a quiet placid beast that likes nothing more than a peaceful life, and no fatalities have been recorded.

Throughout the tropics one comes across a number of large, chunky spiders that are generally known in the United States as 'Tarantulas',

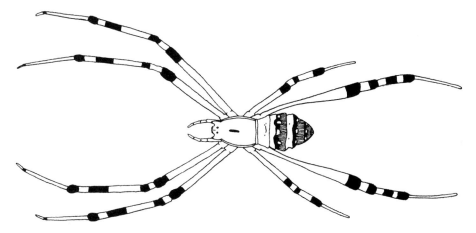

Orb web spiders build typical spiders' webs to trap prey. This species is found in the Far East (Lee Ironside)

and this name seems to be catching on elsewhere. They are not really tarantulas, and should be called, as they were until recently, 'bird-eating spiders'. However, whatever one wishes to call them, though they look alarming they are for the most part of no danger to man. Many of them are reluctant to bite, and even when one suffers such a bite, it is rarely of any consequence. So placid are these animals, not to say beautiful and fascinating, that they are kept by many people. Some keep them because they enjoy studying them, learning from them and breeding them, and others buy them on impulse from pet shops, whence they can be readily obtained these days.

Until fairly recently, the commonest species of tarantula in captivity were the Mexican Red-kneed Tarantula, *Brachypelma smithi*, and the common Pink-toed Tarantula, *Avicularia avicularia*. The latter is still imported regularly from South America, but the Red-kneed has only a limited distribution, and so many were collected over the last few years that there are now restrictions on its collection, and other species are being imported instead. It is still not available, however, since many devoted spider keepers are doing their best to breed these delightful animals. The difficulty is that once males have matured they do not live very long, so it is imperative that keepers in each country advise the holders of local studbooks so that as many matings as possible can be arranged before it is too late. Be all this as it may, 'Tarantulas' were firmly believed to be lethal in the days of the Old West, so much so that it was thought that the only cure was whisky, which came to be known as 'tarantula juice'. The Indians, who tended to be excellent naturalists, had no such mis-conceptions about these spiders; but when the white man appeared they

Emperor Scorpions are large and dramatic but they are infinitely less dangerous than many smaller species (Keith Stiff)

were quick to learn, and soon after they were introduced to this fiery drink more than one of them would take to carrying a spider around with him. At a suitable moment when he was surrounded by enough people he would surreptitiously take out his spider and screech and roll about as though he had just been bitten. Usually some kind soul would come and force a medicinal tot of whisky down his throat before he recovered sufficiently to stagger off to try his stunt elsewhere.

True Tarantulas, *Lycosa narbonensis*, are European spiders that also have a reputation for being lethal. This is not so, though much folklore has been generated about them.

Some of the most attractive of all spiders are the orb web spiders which spin the big web that everybody thinks of as a typical spider's web. There are all sorts of other webs, and orb webs are not of course orbs. They are circular, and very complicated structures, and someone who knows about such things can identify a species of spider by looking at the web. Some interesting research done on orb web spiders some years ago, when drug-taking started to become commonplace, resulted in research into drugs of all sorts becoming more popular. It was discovered that if a spider was fed tiny quantities of a particular drug it would be unable to spin its web in the normal way. In time it became clear that certain types of drug always caused a spider to alter the construction of the web in the same way, which means that a particular aberration in web spinning can be used to identify a given substance.

Most people who come across orb web spiders in their gardens simply do not look at them, but they are truly interesting animals and in the tropics some of them are of a considerable size and flamboyant appearance, such as the Golden Orb Web Spider, *Nephila maculata*, and the St Andrew's Cross Spider, *Argiope versicolor*, both of which are particularly handsome.

Spiders are an astonishingly varied group of animals. There are Bell Spiders, *Argyronera aquatica*, which live beneath the surface of a pond in a little bell of air contained inside a bell-shaped nest; the air is brought down, bubble by laborious bubble from the surface, and renewed or replaced as required. There are spiders that live in burrows with a trapdoor-like lid at the entrance, and there are crab spiders, so-called because of their shape which is somewhat crab-like. One of these in fact seems to be hot favourite at the moment for the title of the most venomous spider in the world. It is the Six-eyed Crab Spider, *Sicarius hahnii*, from South Africa.

SCORPIONS

Though a dislike of spiders seems to be fairly general, scorpions provoke a deep repugnance from people who perhaps have never seen a live one in their lives. Such irrational fear is a great shame since the vast majority of scorpions won't do you any harm at all.

Scorpions really are the most remarkable animals. When the French were testing nuclear weapons in the Sahara, the scorpions were able to withstand the most radiation. They can live through the oddest extremes. A scorpion can survive in the hottest desert, or being frozen in a block of ice. It can do without water for three months, and if it has to it can manage without anything to eat for a year. When it does get a chance to feed it will make a real pig of itself, unlike a spider which will only eat what it needs.

These fascinating arachnids are very sensitive to vibrations. They have hairs on their claws that can detect tiny vibrations in the air, and other organs on the underside of their bodies that pick up vibrations from the ground, and apparently chemical signals as well. It is this awareness of vibrations that is responsible for the fact that they are not easy to see in the wild. As soon as a great big heavy human starts clomping about near them, they vanish into the nearest hole, and you would be astonished at what a tiny crack they can disappear into. On top of all that, scorpions are nocturnal, so one has to be a dedicated scorpion hunter to find any, even in the right environment.

The tropics, the sub-tropics and even parts of the temperate zones are pretty well populated by scorpions, and they turn up in the oddest places sometimes. There is even a colony of them in Essex, just outside London, though it must be said that they were probably introduced inadvertently with container loads of cargo from the continent. Yet, despite all the attractive attributes of scorpions, they are still viewed with loathing. Even the star sign Scorpio is regarded as a symbol of secrecy, lust, darkness and death!

Mating scorpions perform the most intricate and attractive dances together, and mother scorpions carry their hordes of tiny babies on their backs. The youngsters are just like their parent, and some species cart great mobs of them around so that a short distance away it looks as though the female has a woolly coat.

There certainly are scorpions that cause concern, and one of the most venomous is the African Fat-tailed Scorpion, *Androctonus australis*. It is an aggressive, irascible beast that will sting with hardly any provocation. The neurotoxic venom can cause considerable distress and can kill a man in about seven hours if the sting is not treated. The enormous Emperor Scorpion, *Pandinus imperator*, from west Africa is black and looks evil, but is a gentle animal and most reluctant to do any harm. Some of the reports of scorpions stinging people arise because at the end of a night's hunting a scorpion will look for somewhere dark and warm to return to, and where better than an empty pair of shoes that has been left beside a bed overnight. The Trinidad Scorpion, *Tityus trinitatis*, is an animal that has been responsible for a number of human deaths for this very reason. One can hardly blame the scorpion. From its point of view, the diving into an unoccupied shoe is a sensible thing to do.

The largest scorpion in the world is *Heterometrus swannerdami* from India. A male can well be over 18cm (7in) and the largest ever recorded is at the Bombay Natural History Society. It measures 24.7cm (9.75in), and though this is pretty massive it does not mean that it is necessarily dangerous. The already mentioned Fat-tailed Scorpion is not very large, but it is responsible for 80 per cent of all reported stings in Algeria. About a third of the stings from these animals proves fatal. The rest of the victims manage to get to a hospital in time to receive treatment. There is a strange report that in September 1938, 72 stings from this species were recorded in Cairo, while another report from Peshawar on the North West Frontier tells of nine Pathans sleeping in a hut in a nearby village who were stung by some species of *Androctonus*. Eight of them died a short time later. The report says 'a few minutes' though this seems unlikely.

Various species of *Centruroides* such as *noxius*, *suffusus* or *limpidus* can be found in Mexico. Records show that in the years 1940-9, the annual death rate from the stings of these scorpions was between 1,588 and 1,944. In the year 1957-8 there were 2,035 deaths from this cause, most of the victims being youngsters. In the same year in the same area, there were 2,068 deaths attributable to snakebite. Other scorpions of the same genus caused 64 deaths in southern Arizona and New Mexico between 1929 and 1948, a figure which is almost exactly twice the total deaths caused by all other venomous animals together.

The African Gold Scorpion is extremely venomous, but the amount that is injected at each sting is very small so deaths of adults due to this animal are few. The picture is very different, however, when one looks at the incidence for children. In the Middle East about 60 per cent of children stung by these scorpions die as a result. Most stings, in whatever part of the world they occur and from whatever scorpion, are in the sole of the naked foot, and of course in the Middle East most small children run around all the time without any shoes. But even though Gold Scorpions do not cause death in adults, the symptoms are still most unpleasant. The neurotoxic venom causes a tightness of the throat, slurred speech, restlessness, sweating and vomiting, and blueing of the lips and other tissues.

Firms that produce antivenoms are currently listing those suitable for treating the stings of the Fat-tailed Scorpion; the Mediterranean Yellow Scorpion, *Buthus occitanus*; the African Gold Scorpion, *Leirus quinquestriatus*; *Androctonus crassicauda*; *Buthotus saulcyi*, *Hemiscorpius lepturus*; *Mesobuthus eupeua*; *Odontobuthus doriae*, and *Scorpio maurus*.

INSECTS

The only venomous insects of any consequence are ants, bees and wasps. It is true that there are other insects that have venom apparatus, such as the Blister Beetle, *Lytta vesicatoria*, also sometimes called the Spanish Fly though it truly is a beetle. When dried and ground to a powder the resulting irritation is supposed to be aphrodisiac. In fact it can be dangerous. Normally however, unless we go to all the trouble of poisoning ourselves in this way Blister Beetles are of no danger at all. Neither are the bombardier beetles which are able to puff out small clouds of venomous substances to deter small predators.

WASPS, BEES AND HORNETS

As is well known, both wasps and bees of some species will sting. Plenty of others do not and you may not realise that we are surrounded by potter wasps and mud-dauber wasps and many others, since they are no threat at all to us. When they do sting, it is always because we disturb or threaten them. Wasps are generally considered to be of a single species, but the insects that attack your jam sandwiches while you are enjoying a day in the country can be any one of several species, and to tell the difference one has to examine the differences in colouration and other minute details. However, the Common Wasp is *Vespula vulgaris*. Wasp stings differ from bee stings in that those of the former are without barbs, so when a wasp stings, the venom is injected and the sting slides out again like a hypodermic syringe needle. A bee on the other hand has a barbed sting so that it cannot be readily pulled out again. When a bee stings and flies away, it pulls the sting and associated apparatus from its body and leaves it behind. When this happens, the sting should be scraped off the skin with a knife or something similar, rather than be gripped with forceps, which might force yet more venom into the puncture.

As is generally known both wasps and bees have a most detailed, fascinating social life in the wild. Hives, too, are quite incredible. Both natural and artificial living places are essentially a series of individual cells in which are laid the eggs that are to become the next generation, one to a cell. Within these small chambers the eggs grow into larvae, then become pupae, and finally emerge as adult insects. Each hive has a queen that really is the life and soul of the party since without her the hive would die out. Depending on the species of insect, the nest may either be in a hole somewhere, perhaps in a bank, or hanging from something like the branch of a tree. A very few species of bee have been domesticated by man to provide honey, but these bees are never really tame, and may on occasion fly off to start another colony elsewhere.

In an attempt to improve on these strains of domestic honey bees, some beekeepers have tried to cross their insects with other species. This does not always work and an aggressive strain has been the result. In 1956 a number of bees from Tanzania were taken to Brazil. The offspring turned out to be very aggressive and dangerous, and escapees started to spread and colonise the surrounding area most efficiently. They started to move north at an annual rate of 320km (200 miles). It has been calculated that at this rate they should reach the southern part of the USA sometime in 1989, though everybody is hoping that by then the dangerous aspect of the bees' behaviour will have been bred out of them by crossing with other species on the way. Since their original arrival in Brazil, these bees have killed 150 people. They have become so universally detested that the man who originally imported them has been threatened with death on more than one occasion (see also Chapter 10).

Though wasps and bees will leave you alone most of the time, it should be realised that insects cause more human deaths each year than are caused by snakebite. Some people are far more sensitive to the stings of these insects than others and it is almost always those unfortunates who become victims of bee and wasp stings. Because of this sensitivity factor, it can take far more bees to affect one person than another. One unfortunate young Rhodesian made the record books in September 1964 when he was stung 2,243 times by wild bees *Apis adonsonii*. While being chased by the angry swarm he dived into some water and hid for four hours, sticking his head above the surface every so often for a breath of air. When the bees finally left him alone and he went to hospital, more than 2,000 stings were removed from his eyelids, lips, tongue and mouth. Amazingly enough after this dreadful experience he recovered completely.

Hornets, *vespa crabro*, are distinguished from wasps and bees mainly by their larger size. Though they are held in particular dread for their allegedly more aggressive behaviour, they are actually far less likely to sting than the smaller species.

ANTS

Ants can be a real pain in the neck, or the ankle, or just about anywhere. Until one has encountered a determined attack by ants it is difficult to imagine just how painful they can be, and it is not always the biggest that cause most damage. There is a very small species of red South American ant that is especially painful. It is generally conceded, though, that the most dangerous ant in the world is the Black Bulldog Ant, *Myrmecia forficata*, from Australia and Tasmania. It uses jaws and sting at the same time when it attacks. Not all species have stings, but even those without them are capable of squirting formic acid from the tip of their abdomen at predators. Allied to this they have powerful jaws that can give a painful bite, at the same time making an incision into which formic acid can enter. Other species of fire ant of the subfamily *Myrmercinae* from various parts of South America are also generally regarded as pretty nasty.

Ants as a whole are a most rewarding group of animals to study for their complex social life styles. Many people will be familiar with the fact that ants milk aphids for the sweet secretions they produce, tending their own particular herds of these insects and protecting them from parasites. Many species of ant are involved in one way or another with other animals, and some of them even practise slavery. *Formica sanguinea,* for example, raids the nests of other species in order to satisfy its need for slaves. Pupae are brought back to the nest and tended until they emerge as adults, whereupon they work thereafter for the *sanguinea* ants.

CENTIPEDES

Centipedes and millipedes are completely different animals. Millipedes are vegetarian and since their food is therefore not very likely to run away, they are not very fast-moving animals. Each segment of the body has two pairs of legs. Centipedes on the other hand are carnivorous. They have one pair of legs per segment, and they can move at astonishing speeds to enable them to catch fast-moving prey.

Small centipedes from temperate regions are not a threat to man, but some of the large, tropical species can cause much pain by injecting venom through claws which are modified legs on the first body segment of the animal. The venom is used primarily for subduing prey animals. Many tropical species are brightly coloured and the biggest by far is the Giant Centipede, *Scolopendra gigas*, from tropical America and the West Indies. Centipedes the world over are generally to be found in loose soil, in leaf litter and under stones, and though they are commonly known as centipedes they do not necessarily have a hundred feet. Some of the species in the Far East are used by anglers in that part of the world to bait their hooks and can be bought for this purpose from angling shops. The two species *Scolopendra morsitans* and *Scolopendra subsnipes* are both used in this way.

MOLLUSCS

CONE SHELLS

Cone Shells such as *Conus geographus* are molluscs that live primarily in the Indo-Pacific seas of the world, on and around the coral reefs. They have a device much like a harpoon that they can fire from the slit-like opening in the shell. The harpoon is barbed, and is used both to obtain food and to defend the animal against a predator. Normally they do not pose any sort of a problem and accidents only occur when someone such as a diver or a shell collector kicks or handles the animal. The mottled brown shells are much sought after by shell collectors. The neurotoxin that Cone Shells secrete is very dangerous and is said to be able to kill in fifteen minutes.

OCTOPUSES AND SQUIDS

Octopuses and squids may seem to be unlikely creatures to be in a book about venomous animals, but at least one species of Flying Squid, *Onychoteuthis banksi*, has a toxic bite. Despite this it is sold in markets throughout its range, as food. It lives at a depth of 500m (1,650ft) during the day and only comes to the surface at night when hardly anyone is about; consequently the only people at risk from these animals are the fishermen who catch them.

The octopus that is most notorious for the toxicity of its venom is the Blue-ringed Octopus *Hapalochlaena maculosa*. It occurs in shallow waters around the coasts of Australia, where it feeds on molluscs and crabs which it opens with its strong beak. It is a tiny animal, only 10cm (4in) long, but is quite deadly to man, and has been responsible for numbers of excruciating deaths. The pain is severe and death supervenes in a very few minutes. This octopus is common under rocks at low tide, which is why it is frequently disturbed by holidaymakers.

COELENTERATES

SEA ANENOMES

Sea anemones are all capable of introducing venom into the small marine creatures on which they feed. If one runs a finger through all the tentacles of some species, a faint sensation of tingling as a result of the venom may be felt, but that is usually all. The most toxic sea anemone would seem to be the Matamulu, *Rhodactis howesii*, from Samoa though the danger does not exist from being stung, rather from eating it raw, which is done by local people. This can cause the death of the diner due to respiratory failure.

JELLYFISH

Jellyfish too are venomous, though the injuries caused by most of them are grossly exaggerated. In Britain each year there are many jellyfish scares due to the Moon Jelly, *Aurelia aurita*. Swimmers rush from the water in a panic whenever this animal is sighted in the waters off the beach, but in fact it is a most inoffensive beast. Nonetheless stories persist about the stings that it is alleged to have caused. Some years ago a doctor at a popular seaside resort became philosophical about the numbers of patients that came to her each summer suffering from what they considered to be near-fatal jellyfish stings. At weekends some of these people would arrive at her house when she was off duty. Each time she would go into a spiel about rest and how if the patient took it easy there would be no lasting damage, and she would end by giving a placebo. She discovered that unless she did this the patients would not believe the harmlessness of this jellyfish, and would go away and develop a whole variety of psychosomatic symptoms.

At the time the doctor had a small child. He became so used to hearing what was said on each of these visits that it occasionally happened that the doctor was at the end of the garden and unaware that anyone had come to the door. The child would sometimes open it and on being told what the matter was, would go into the same reassuring routine as his mother. Whenever this happened, the patient apparently recovered just as well as when the doctor treated him. It is strange how people panic. A few nettle stings do not cause concern to anyone, but a similar discomfort caused by a jellyfish is regarded as a prelude to certain death.

Some species of jellyfish are more dangerous. The much-feared Portuguese Man O' War, *Physalia physalis*, causes stings like cigarette burns, but these are extremely unlikely to be fatal unless a person suffers a great many of them. Generally the sensation has disappeared a day later. The float of this species looks more like a cheap plastic toy rather than an animal. The dangerous parts of the animal are the dangling tentacles, which can be many yards long. With all jellyfish stings that cause minor discomfort, alleviation can be achieved by rubbing the spot instantly with meat tenderiser.

One group of jellyfish venomous enough to kill human beings is collectively known as box jellyfish from their box-like shape. The Sea Wasp, *Chironex fleckeri*, is unarguably the worst though several relatives come a close second. This species is known to have been responsible for over fifty deaths, which occurred usually between three and ten minutes after a sting, preceded by sweating, convulsions, blindness and respiratory paralysis. The venom is so toxic that in laboratory experiments a solution diluted 10,000 times caused the death of animals even before the syringes could be extracted. A large Sea Wasp may have up to 61m (200ft) of tentacles; contact with only 6m (20ft) is sufficient to cause death. Victims of this species in Australia who have died as a result of being stung have been found to have considerable quantities of frothy mucous in their lungs and air passages. Even mild stings are unpleasant; they cause an acute burning sensation and can result in weals that remain for months. Sea Wasps sometimes are found very close to the shore, and one Australian girl of nine who was stung was standing in only 76cm

(30in) of water at the time. Funnily enough the only protection against the stings of these animals is the wearing of women's tights, which the stings cannot penetrate. Because of this they are regulation apparel for the life-saving teams in Queensland.

CONCLUSION

Some other animals could, strictly speaking, be included in this book, but the effect of their venom is of no consequence to man. All the corals, for example, fall into this category and some of the sea urchins and starfish. So too do the hydra, tiny freshwater organisms that are capable of shooting out venomous darts. But if one touches them the result is so inconsequential that no effect at all can be felt.

While we have seen that many animals in the world are venomous to a degree, it should be remembered that most are not, and of those that are only a comparatively small number of species are of any danger to man. Even these are reluctant to bite or sting unless they are disturbed or interfered with, as they would far rather get on with living without us. It should also be realised that most victims of venomous animals survive; only a very few deaths are caused due to the fact that an attack is almost always very fast and the animal does not have time to inject an appreciable amount of venom.

2 · Venoms and Antivenoms

Suffering from the bites and stings of venomous animals is so horrendously unpleasant that for centuries man has been trying to find a cure of some kind, and all sorts of wonderful attempts have been made to alleviate the injuries and deaths. By themselves, most of these cures are pretty useless and some of them are actually harmful. However, given the nature of the subject, it is not surprising that many of them appear to work, and are therefore to be believed. Most people who are bitten or stung show little or no effect and of those that do, taken overall, very few die. One comes across the oddest contradictions. It is not uncommon to hear of two virtually identical cases where one person had died and the other had suffered no ill effects from, say, the bite of a snake.

Because of this confusion some ineffectual cures have gained a reputation for success. A person who believes himself to be the victim of an attack by a venomous animal may suffer only a little or not at all for one of several reasons. The commonest is that the culprit may not be venomous anyway. In captivity it is not usually difficult to distinguish between a venomous species and one that is not. In the wild it is another matter as very often one only gets a quick look at an animal before it disappears down a hole, and it is very easy to confuse one species with another. Doctors who treat patients for snakebite can all tell stories of people who have been bitten by harmless snakes, but who have apparently developed typical symptoms of snake poisoning, usually due to fear and sometimes as a result of an infection occurring at the site of the bite. Even if the bite is dangerous and painful the greater part of the world's venomous animals, as we have seen, are not lethal as far as man is concerned, so although there may be varying degrees of reaction in the bitten person, death may not result. Any bite or sting may result in only a small quantity of venom being introduced into a wound, sometimes through only one fang. Clothing

(right) *This Wolf Spider is venomous but cannot possibly hurt man*

too may inhibit a lethal dose, and the fact that the animal has perhaps already used up much of its available stock of venom could mean that there is not much left at the moment of the strike. If any one of these things happens when a person is bitten, any subsequent treatment will be acknowledged to be successful if the victim does not die.

Over the years there have been as many treatments for snakebite and other venomous injuries as there are species of animal that can inflict them. Alcohol seems to be the most common, not to say popular, medicine to use. The astonishing thing is the quantities of the stuff that were recommended for cases of snakebite in the days of the old West. During the American Civil War it was thought essential to pour a gallon of whisky into the sufferer. At that time it was worth $450 a gallon, and one poor old army quartermaster complained that it was cheaper to replace the men than the whisky. Another American remedy for rattlesnake bite at the time suggested $1/_2$pt of bourbon every 5 minutes until 1 quart had been consumed. About a hundred years ago the recipe was 1qt of brandy and $1^1/_2$ gallons of whisky within 36 hours. It was astonishing that anybody survived this sort of treatment. It cannot in any case have done the patient much good since it must have put an additional strain on an already suffering liver, not to mention the kidneys and the circulatory system.

A traditional and oft-quoted remedy was to cut several deep incisions across the site of the bite or sting and let the blood flow freely from it. This caused considerable damage to the affected part and did almost nothing towards stopping the venom from entering the bloodstream. If such treatment is to have any effect at all it has to be performed so soon after the attack that by the time a patient has reacted and hobbled to a stone to sit down and draw his knife it is already too late. The other traditional course of treatment involved someone sucking the venom from the wound, and if this is done really promptly it can have a beneficial effect on the patient. The only problem is that there are not many people in the world who do not have a bad tooth or gum disease or a mouth ulcer. All these conditions enable the venom to enter the sucker's own bloodstream which can result in the death of two people instead of one. For some reason, both these courses of action were usually associated with the application of potassium permanganate, which was useless anyway, and if used in any quantity was terribly destructive to the tissues of the bitten person.

(left) *Every child knows about Spiderman who acquired his powers after a bite from an irradiated spider!*

A spectacular remedy for snakebite that one sees performed in Westerns by macho cowboys is the amputation of the hurt finger with a large, unhygienic-looking machete, or in some cases the finger is even blown off with a trusty six-shooter while the hero flinches slightly. Such action might well stop the absorption of the venom throughout the system, but it has been shown that a dog absorbs venom from a snake bite in about one minute, so the hero would have to be very quick, and would probably do better to chop his arm off at the shoulder than bother about the bitten digit.

However, even though most of these solutions could not work at least one can see the thinking behind them, namely to remove the venom or prevent it from travelling throughout the body. Some of the other methods of achieving the same ends sound bizarre in the extreme. Soaking a bitten hand in paraffin is supposed to be good or, even better, splitting a chicken in two and wrapping the two halves round the damaged part before covering the whole ungainly lump with bandage until the chicken carcase has sucked the venom from the wound, whereupon it should be burnt. If all else failed a friend of the victim might try what was supposed to be an infallible Indian cure for snakebite. According to the instructions, the friend was supposed to sprinkle cold water on the face of the victim as he lay suffering on the ground and say to him three times in a commanding tone of voice, 'Get Up! It is the command of T.C. Ramachandra Rao'.

Written remedies for snakebite go back a very long way. In the writings of Nicander, second century BC, one can discover a remedy that is still suggested in some places today. He said that one ought to squeeze toad urine into the wound. When every last drop had been wrung out of the poor old toad, it should be killed and the wound left to soak in the liquid. What is interesting is the fact that the toad takes no further part in the proceedings. One would think that since specific instructions are given to kill it, the next sentence would advocate doing something else with it, as for example in the split-chicken remedy. Other cures that persist to this day can be found in the writings of Celsus in the first century AD, and of Galen who lived from AD 130-200.

All over the world people who handle and are frequently bitten by snakes claim to be immune from their bites, and though there is evidence that some people do develop a certain immunity, many have died as an immediate result of the next snakebite. There used to be an old snake catcher many years ago in West Bengal who supplied reptiles to all the animal dealers. Ibrahim was his name and he had certainly been bitten on numerous occasions, usually by cobras. He claimed to be

Johannes F. Schuhknecht Jr.

"Quality Venoms for Medical Research"
P.O. Box 252
Staatsburg, New York 12580   U.S.A.

Prices are quoted in U.S. dollars.   Shipments FOB Staatsburg, New York.
All snake venoms are lyophilized.   Inquire for unlisted species.
Price list effective: January 1, 1987

SPECIE	100 mg	250 mg	500 mg	1 gram
Agkistrodon c. contortrix	9.	21.	37.	65.
Agkistrodon c. mokason	9.	21.	35.	60.
Agkistrodon p. piscivorus	5.	12.	21.	35.
Bitis arietans	12.	28.	45.	80.
Bothrops asper	15.	34.	60.	115.
Bothrops atrox	12.	28.	47.	85.
Bungarus caeruleus	65.	160.	310.	600.
Bungarus fasciatus	15.	35.	55.	100.
Bungarus multicinctus	85.	210.	410.	800.
Crotalus adamanteus	8.	18.	32.	55.
Crotalus atrox	8.	15.	27.	45.
Crotalus h. atricaudatus	9.	21.	35.	60.
Crotalus h. horridus	9.	21.	35.	60.
Crotalus s. scutellatus	25.	60.	110.	200.
Crotalus v. viridis	15.	35.	55.	100.
Dendroaspis polylepis	35.	85.	165.	300.
Micrurus fulvius	90.	207.	385.	750.
Naja h. haje	9.	21.	35.	60.
Naja melanoleuca	14.	32.	50.	90.
Naja n. kaothia	6.	14.	24.	40.
Naja n. siamensis	6.	14.	24.	40.
Naja nigricollis	9.	21.	35.	60.
Naja nivea	10.	24.	40.	70.
Ophiophagus hannah	12.	28.	45.	80.
Sistrurus miliarius barbouri	12.	28.	45.	80.
Vipera r. russelli	15.	35.	55.	100.
Bitis gabonica	15.	34.	60.	115.
BEADED LIZARDS				
Heloderma horridum	35.	85.	165.	300.
Heloderma suspectum	65.	160.	310.	600.

The milking of snakes is a chancy, dangerous business, and the cost of snake venom reflects this (John Nichol)

immune but eventually died as a direct result of a cobra bite. Even where there is a degree of immunity present in a person, later bites by a snake of the same species still cause considerable trauma and distress. George Hemsley, the former founder of the well-known rattlesnake religious cult in the south-east of the United States, claimed to be immune and said that he had survived 400 earlier bites. He was finally killed by the bite of a rattlesnake. William E. Haast, who at one time was the director of the Miami Serpentarium, was bitten more than a hundred times by a variety of venomous snakes during his lifetime. Research showed that he did have a degree of immunity, but he also suffered some serious bites.

The breakthrough in effective treatment for snakebite only began to appear in 1887 when an American named Henry Sewell began to wonder if an animal could become immune to animal venoms if tiny, sub-lethal doses were injected. He decided to do research into the subject and, in Michigan where he lived, began to inject pigeons with snake venoms. He lost an awful lot of pigeons, and though he continued with his work the next useful stage in the story was reached in 1895 when a French bacteriologist working in the Pasteur Institute in Paris managed to develop the first antitoxic serum. The bacteriologist's name was Albert Calmette, and by the time he died in 1933 the whole business of producing antivenoms was well under way, and today forty laboratories around the world produce a wide variety of antivenoms for use against snakebite, scorpion stings, envenomation by Sea Wasps and other animals besides.

Yet even today, sophisticated though modern antivenoms are, it is still a risky business using them. They have saved the lives of countless people, but it is important to use them only in cases of extreme urgency. In the event of a bite or sting by an animal that is very unlikely to cause death, a variety of first-aid measures and treatments are available that are more effective. If antivenom is used in minor cases the effects of the treatment can cause more problems than the original envenomation. The problem is that antivenoms are made using horse serum, and some people are allergic to it. On top of that one often has to inject great amounts of the stuff, and in an allergic person this can cause collapse. So even when antivenom is used, a small injection of the substance is tried first, to test for reaction. Before such a test a syringe of adrenalin has to be prepared because if the patient is sensitive to horse serum collapse can happen very fast. If the case is serious enough to merit the use of antivenom and the patient is reacting to the test injection, the only practical possibility may be to give a blood or plasma transfusion. Even when a sensitive person does not collapse, serum sickness may lead to fever as well as an itching sensation and sore joints which can still be felt ten days afterwards.

MANUFACTURE OF ANTIVENOM

The manufacture of antivenom is a long, complex process. The story begins in those parts of the world where the snakes, the scorpions and the spiders live, against whose bites and stings we need antivenoms. Specimens are usually collected in the field by local hunters who are paid for each specimen they bring into the laboratory. The difficulty with this method of doing things is that one may not be able to get hold of enough specimens of a particular animal to meet the demand. The appearance of many animals in a particular area in the wild is frequently seasonal; or it may not be possible for trappers to get to an area and back through floods or similar obstacles, and if the hunter with a load of animals is held up somewhere some of his stock may die before it ever gets to the lab. So to try and establish some sort of control over these disadvantages a few of the more forward-thinking establishments are now beginning to breed their own stock of animals. However, this system is still in its infancy and it will probably be quite a few years before more than a tiny percentage of the animals required are bred in captivity. So the hunters continue to go out to capture the live animals, usually on a freelance basis, though in one or two places the system is rather better organised. In India for example, there is a snake catchers' co-operative at six Irula villages.

In many parts of the tropics one tends to find that there are whole communities who have traditionally made their living by trapping wild animals. Sometimes these collectors are specialists who deal only in, say, reptiles or birds; in other cases they are hunters of anything they know they can sell. In India there is an ancient tradition of snake catching and charming in Rajasthan, though there are other snake-collecting communities all over the country such as the Santal tribals in West Bengal. But wherever the snake catchers come from, they are all superb field naturalists and it is almost unheard of for them not to bring in those numbers and species of snake for which they have been asked. For many years only snake antivenoms were produced since it was these reptiles that were responsible for most human casualties. Nowadays other antivenoms are also being made.

Sometimes the laboratories are proper medical establishments attached to hospitals. At other places they are commercial organisations that make their profits by supplying antivenoms to hospitals, and sometimes they are small firms that collect and prepare the venoms only, then sell them to other people who have the facilities to make antivenom. The preparation of the final product may involve several sophisticated stages, but basically the method used to make antivenom is standard. The animals received from the hunters are kept in enclosures or containers of some sort until they are required. Invariably each species is kept on its own. When venom is wanted an animal is caught, and if it is a snake, it is usually held by the back of the head and made to bite through a membrane covering the top of a small container. The snake is helped to eject its venom by gentle pressure on the venom glands. When they are empty the snake is returned to its enclosure. In time it will produce more venom and the business will be repeated. The stress caused to the animals by this treatment is considerable and they don't often last long.

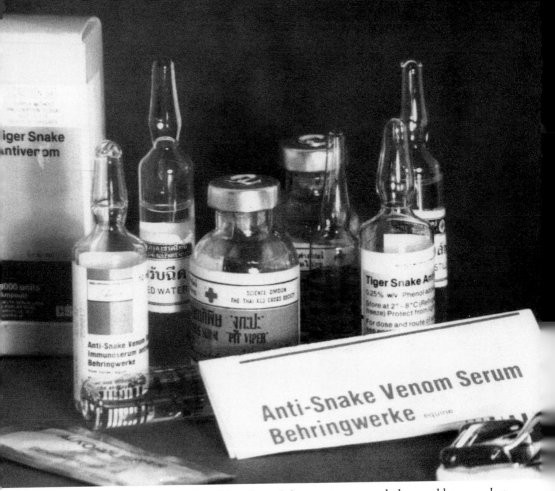

Antivenoms are produced by thirty laboratories around the world to combat the bites and stings of the most dangerous species of animal (John Nichol)

Sometimes this milking of the snakes for their venom is done in public as part of a demonstration. People the world over are fascinated by any sort of dangerous animal, and such snake shows attract enormous interest.

Some animals cannot be milked of their venom in the conventional fashion. Such species, like the Gila Monsters and the rear-fanged snakes, are induced instead to chew on a piece of plastic or rubber, and the liquid is rinsed out with sterilised water, or picked up from the surface with a capillary tube, or even with filter paper.

Before venon is extracted from particularly aggressive individual animals, they may have to be chilled slightly or even anaesthetised to render them manageable. Some species of venomous animals are stimulated to produce their venom by applying electrical shocks, while from others the venom cannot be obtained unless the animal is killed.

When sufficient venom, which is a yellowish, slightly cloudy liquid, has been collected from a single species of snake, the container is placed in a centrifuge to separate the venom itself from any debris that it may contain, such as dead cells. The resulting liquid is perfectly clear. This is freeze dried and scraped off the plate, at which time it looks like small yellow needles or crystals. This dried venom is now ready for making into antivenom.

Labs try to freeze liquid venom after about six samples have been obtained until there is a sufficient quantity to put into the freeze drier, which is done without thawing. Once it has been freeze dried, it can be stored away from light and dampness for years without losing potency, but heat causes venom to become inactive. At 65°C (149°F) and over, it begins to lose its potency, and if kept at 80°C (176°F) for five minutes it ends up almost totally harmless though there are exceptions - as there are to almost every single fact about venomous bites and stings! Wagler's Pit Viper venom is unique in that it will remain as dangerous as ever even after spending a quarter of an hour at a temperature of 125°C (257°F). The venom of the Gila Monster is also fairly resistant to heat. Ultra-violet light and X-rays also cause a solution of snake venom to become inactivated.

The extraction of venom is a slow process. It takes ten Timber Rattlesnakes, for example, to provide about a teaspoonful of liquid venom, and only a tiny part of this remains when it has been centrifuged and dried. The liquid is slightly viscid, and has almost no discernible smell. Its taste is said to be slightly sweet and weakly astringent, and it is a little acidic, having a pH of about 5.9. The venom from these ten rattlesnakes provides only about 1.35g (1.04oz) of freeze-dried material.

Dried venom is expensive stuff, and often people who are used to handling venomous animals feel that they would like to make themselves large sums of money by selling venoms. When they discover the amount of animals one needs to obtain a commercial amount of liquid, and the preparation involved, they usually, and quite rightly, lose interest.

At the time of writing, typical prices for freeze-dried venoms are as follows, in $US per 100mg:

Snakes		Toads	
Bitis arietans	$10	Bufo bufo	$20
Vipera berus	$150		
Crotalus atrox	$8	Scorpions	
Dendroaspis polylepis	$50	Androctonus australis	$330
Naja naja	$10	Buthus occitans	$550

Ants		Hornets	
Pogonomyrmex wheeleri	$3,200	Vespa tropica	$3,000

Honey Bees	
Apis mellifera	$700

Once these venoms have been prepared or bought from suppliers, the antivenom has to be manufactured from them. A solution of the venom is injected into a horse, the dosage being between one-tenth and one-hundredth of a lethal dose. A further increased dose is administered at weekly intervals for about three months, by which time the horse has built up an immunity to that specific venom. By the end of the three months the horse will have received several doses that would have normally proved lethal had the smaller doses beforehand not enabled it to become immune. By now the blood of the horse is rich in antibodies and immunoglobulin proteins that neutralise the venom with which it has been injected. Blood is then taken from the horse, and separated. The serum thus obtained has now become antivenom. If the horse is given periodic booster injections it can be used for six or seven years, during which it may have given 700 litres (154gal) of blood. The basic serum is further treated to purify it, enable it to be stored effectively, and to standardise its effectiveness; for although there are various recommendations, there is no international standard for antivenoms.

ANALYSIS OF VENOM

Animal venoms are not all of one type. Each has its own cocktail of ingredients that have different effects upon a victim, and as a result we need suitable antivenoms for each type of venom. It is not much use treating someone who has been bitten by a cobra with an antivenom for a rattlesnake, though it is true that there are preparations that may be used to treat the bites or stings of several animals. Such polyvalent antivenoms are not as good as one that is monovalent, but in many cases they are most useful. It could be that a patient is brought into hospital suffering from snakebite from an unidentified species, so an injection of a polyvalent serum may be the only option. Nowadays there are monovalent serums available for all the dangerous snakes.

Animal venoms are remarkably complicated mixtures of proteins. At one time there were all sorts of guesses as to what venoms were made from and which part of the body was responsible for their production. Gila Monsters were at one time said not have an anus, so it was thought that venom was the result of putrefying food in the gut. It was a physician in America, S. Weir Mitchell, and an English biochemist, Norris Wolfenden, who first established that venoms were a mixture of proteins, and John de Lacerda, a French physician working in Brazil in 1880, discovered the presence of enzymes in snake venom. Nowadays twenty-five of them are known to be present. Each has a specific function. Hyaluronidase promotes the rapid spread of venom from the site of the puncture, phospholipase breaks down the cell membranes of the victim, and protease liquifies the tissues. While the nucleotidases are not toxic they help the animal digest the victim once it starts to eat it, for venom is really a highly specialised saliva which not only enables the snake or the spider to capture and kill prey efficiently, but also to consume the tissues faster after they have been partly predigested by the venom.

Most animal venoms fall into two groups. Some, produced by the cobras and related snakes, together with various other animals such as many of the invertebrates, are neurotoxic, which means that they affect the nervous system and kill a victim ultimately by stopping the action of the heart and lungs. Actual damage to tissues is not too great. On the other hand, the venom of the vipers and pit vipers is haemotoxic, which means that it breaks down the tissues of the victims. The walls of blood vessels are destroyed and as a result blood seeps out beneath the skin, appearing to a watcher as enormous bruises or, in more extreme cases, as a blackening patch beneath the skin. Swelling and pain are considerable. Each type of venom, however, contains elements of the other, and all of them consist of a cocktail made up from varying proportions of the following eight types of ingredients:

1 Neurotoxins, which act on the central nervous system
2 Haemorrhagins that destroy the walls of the blood vessels
3 Thrombose to produce blood clots
4 Haemolysins which destroy red blood cells
5 Cytolysins which act on leucocytes and other cells
6 Anticoagulants to prevent the blood clotting
7 Antibacterial substances
8 Ferments that prepare the food for digestion

Elapid snakes, which are the cobras and their relatives, have venom that is rich in elements 1, 4 and 6, while vipers and pit vipers venoms contain a predominance of 2, 3 and 5. Both types have a certain amount of 7 and 8 as well.

As if all this were not complicated enough, venoms are often fairly specific in that they will act far more quickly on one type of animal than another. Sea-snake venom, for example, works faster on eels and other fish than on other types of animals.

As may be supposed, considerable research has gone into snake venoms, far more than into the venoms of other species, though work on spiders and other invertebrates is being done more and more these days. The secretion of venom begins before the animal is born, and baby snakes are every bit as dangerous to man as an adult of the same species. The liquid is manufactured in the main venom gland, which is situated just above the mouth of the snake. It is stored in a reservoir until it is needed, at which time it travels down a duct to the fang. Just before it reaches the fang there is an accessory gland that adds elements to the venom; which make it even more dangerous, just before it is ejected by the animal into the tissue of the victim, by muscular pressure on the reservoir. A snake, or any other venomous animal for that matter, does not need to use all the stored venom in a single bite, and in fact does not do so. With the potential to kill several animals the size of a man, it is pointless using more than a tiny fraction of that venom on a small rodent, or even a predator. There is usually no need for the predator to die, and in fact death may be counter productive since a dead animal cannot learn, nor pass on the learning to other members of its species, not to attack that snake a second time.

Once the glands and reservoirs are empty of venom a snake has to wait some little time before they are filled once more. The rate at which this happens depends on various factors, especially the ambient temperature, and more is produced at a faster rate in warm weather than when it is cold. Tests showed that in summer the Palestine Viper needed sixteen days to refill the reservoir, while rattlesnakes took fifty-four. This delay is obviously of importance to snake farms that keep their animals for venom production, and it is generally recommended that no snake is milked at less than three-week intervals. It has also been noted that the amount of venom available at each milking declines, though this might be because of the abnormal stress on a snake kept in these conditions, rather than a true reflection of what happens in the wild.

In experiments it has been shown that a snake almost never injects more than about half the available venom in a single bite, and the amount used is generally far less than this, usually not more than 11 per cent on average. In a series of tests, one snake struck at a target 22 times in succession, injecting venom on each occasion. It must have been an exceptionally dim snake. They will usually learn a lot faster that they are getting nowhere in this sort of situation and give up in disgust. More typically, another snake in the same series of tests made 10 consecutive strikes, injecting venom in only 4 of those, before it refused to continue with the experiment. It was then milked to discover how much venom was left, and the amount was found to be 85 per cent of the total.

Animal venom is often so startlingly toxic that it is easy to forget what small quantities each animal actually produces. As a rule vipers manufacture more venom than other types of snakes, but even vipers do not produce that much. Snake farms and laboratories producing snake venom are careful to keep records, and it has been found that comparisons can be made of the amount produced by each species, though results with another group of identical snakes might produce an entirely different set of figures. But for what it is worth, a Jararacussu can produce about 1,500mg in its lifetime at a snake farm, a Cottonmouth 1,000mg, a Western Diamondback Rattlesnake 1,100mg. Old World viperids give far lower yields: a Puff Adder provides 700mg, a Gaboon Viper substantially more at about 900mg, and cobras rather less than Puff Adders. In a single milking it is rare to obtain more than 5mg from a snake, though it is on record that an Eastern Diamondback gave 6.25mg on one occasion.

It is remarkable that the blood of some snakes is toxic when it is injected into mammals, and it has been suggested that the venom glands filter and perhaps concentrate compounds from other parts of the snake's body as a basis for the venom. But careful analysis has demonstrated that none of the toxic proteins that are the major constituents of venom are to be found elsewhere in the body, including the blood. Furthermore, if toxic blood from venomous animals is injected into other animals, and then injections of antivenom given, the latter does not neutralise the effect of the former.

The preparation of synthetic antivenom would seem to be a worthwhile field of research, given the dangerous and unpredictable nature of obtaining antivenom from animals, but as yet it has proved impossible to artificially manufacture the large, complex molecules of the fifteen to twenty active substances that make up the major part of animal venom.

At one time it was thought that if someone were bitten by, say, a cobra in India, cobra antivenom, pure and simple, would help the victim recover. This was fine when the nearest place in India to obtain Indian Cobra antivenom was somewhere like the Haffkine Institute in Bombay, but once air travel became the normal method of transportation, companies in one country were able to offer their products to other countries, and animals from one country began to be kept more commonly elsewhere, making it necessary to keep stocks of appropriate antivenoms. As soon

Giant centipedes from the tropics can give a most painful bite (Herbert Lieske)

as this happened, it became obvious that if a person was bitten or stung by an animal, it was no longer sufficient that a monovalent antivenom relevant to that animal was applied. It had to be antivenom obtained from an animal that came from the same part of the world as that of the snake which caused the injury in the first place. So, a bite from a cobra in India could not be effectively treated with a venom from a cobra of the same species from Thailand or Malaysia. Nowadays, antivenom laboratories keep stocks from each geographical area.

Two questions that are invariably asked in conversations about snake venoms refer to their relative toxicity and the effect they have on man. Bearing in mind that definitive answers within this field are as rare as hens' teeth, it has been estimated that the venom of the Asian Cobra can be compared in this way with other toxic substances:

- It is 40 times as toxic as sodium cyanide.
- It is twice as toxic as strychnine.
- It is 7 times as toxic as the poison of the Fly Agaric toadstool, which is the red one with white spots that pixies sit on.
- It is about the same as the venom of the Yellow Scorpion of the Middle East, and about 5 times as toxic as the venom of the Black Widow spider.

Having said all that, it should not be forgotten that between a third

The world famous Butantan Institute in Brazil where all of South America's anti-venom is produced. In the foreground are the snake enclosures (Herbert Lieske)

and a half of all bites from venomous snakes result in little or no venom being injected.

SYMPTOMS OF BITES

Bearing in mind the different formulas of the various types of venom, it is not surprising that one hears of an apparently endless list of unrelated symptoms appearing in victims of snakebite. This is undoubtedly made more confusing due to the misidentification of the animal that caused the injury, as well as secondary infections, gangrene and psychosomatic symptoms. However, what typically happens with cobra bites, as well as with the bites and stings of other animals that inject neurotoxic venoms, is that there is little or no sign of envenomation for fifteen to thirty minutes after the bite. Then a degree of pain at the site of the puncture begins to be felt. There can be no doubt that it is painful, but most victims do not claim that is it unbearably agonising. Thereafter a patient begins to complain of heavy or drooping eyelids, which is followed by difficulty with swallowing and speaking. At this point patients can no longer control the swallowing reflex and begin to drool saliva. Dizziness follows and sometimes nausea, together with a feeling of general weakness. If the pulse is felt at this stage, it is irregular. The unfortunate victim starts to gasp as the venom begins to stop the lungs working, and if he is going to die, he will do so soon after, either because he can no longer breathe or because his heart stops. But most victims recover, often after a few hours, and rarely are there any lasting effects. Patients who have survived being bitten by cobras often report that at the time they were aware of what was happening and they felt no anxiety nor real pain.

Bites from other elapids often result in no symptoms for up to six hours after the bite, but once they have started they progress very quickly. Sometimes patients complain of what feels like a great weight on their chest when they are trying, with difficulty, to inflate the lungs. By contrast, bites from coral snakes and kraits are apparently more painful, and victims talk of shooting pains.

Rattlesnake and viper bites generally are another kettle of fish altogether, and most are certainly not a pleasant way to die. The site of the bite is instantly very painful, a pain which some people describe as if inflicted with a hot needle. If the site then goes numb, this is usually a sign that the bite is a serious one and that a considerable quantity of venom has been injected. If a victim is lucky, shock will supervene before much else happens and this will spare the patient the intense agony which he might otherwise suffer. Otherwise the mouth starts to tingle and the bitten part begins to get larger. Viper bites are often in the foot or ankle, and the lower leg swells until it is of a greater diameter than the thigh. The affected limb becomes so tender that the presence of clothes are very painful and even a simple cover such as a sheet may not be borne without pain, and a slight draught or breath of air has caused patients to scream. Vomiting and spasms are the next symptoms and, as the venom causes the breakdown of blood vessels and other tissues, first the limb and then often other parts of the body begin to discolour until the whole area looks like an enormous bruise which can sometimes affect most of one side of the body.

Sufferers complain of intense, blinding headaches, and indeed blindness. The kidneys can become seriously damaged, and urine is discoloured with blood. Bleeding from the gums and sometimes the nose and ears is not uncommon. Intense pain is always present, and if a person is going to die, he will ultimately bleed to death internally as all the working parts of the body leak blood. If a patient survives, extensive scarring often results where there has been extensive necrosis, and gangrene and subsequent amputation of affected limbs are fairly common in victims

The Adder is the only British venomous snake, and despite its reputation it has caused very few deaths (Herbert Lieske)

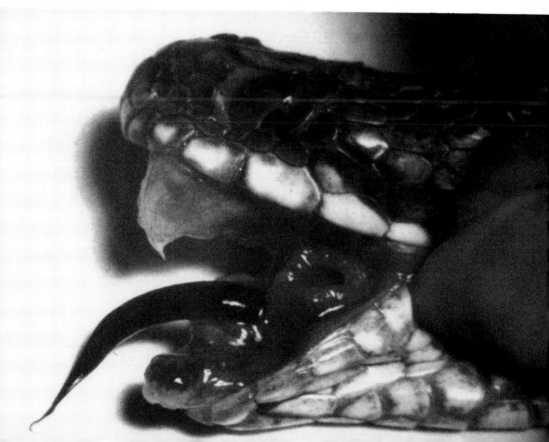

bitten by snakes of this type. In Florida, 17 per cent of all amputations on children are as a direct result of snakebite. A fortunate patient can sometimes lose consciousness after a quarter of an hour, but due to the unpredictable nature of snakebite, victims of viper bites can last up to twelve days before ultimately dying of cerebral haemorrhage.

Gila Monster bites cause sweating, nausea and thirst, followed by a sore throat and a ringing in the ears. Next weakness and rapid breathing results in a feeling of faintness, and then in collapse. The bite of one of these lizards is a painful business as their venom contains substances called serotonin, which is there specifically to produce pain, in order, presumably, to dissuade survivors from ever handling a Gila Monster again.

All these horrific symptoms of snakebite and lizard bite refer to man and are caused when venom enters the bloodstream. Although it was said earlier that taking venom by mouth is not too dangerous unless one has bad teeth, gum diseases, ulcers, or scratches on the wall of the gut (and that includes most of us), large doses of animal venom by mouth can be toxic. To kill oneself by drinking rattlesnake venom one would need to drink 750,000 times the normally injected dose, though a lethal cobra dose by mouth would be only about 100 times the normal dose. Man has been investigating animal venoms since the beginning of time. The Roman poet Lucan, said:

Mixed with the blood, the serpent's poison kills;
The bite conveys it. Death lurks in the teeth.
Swallowing it worked no harm.

Though the effects of snake bite as described are horrific, it is important not to get the matter out of perspective. It has been estimated that half to a third of all snake bites in the United States are caused by handling venomous reptiles, and in Australia, which is renowned for the number of venomous snakes, there are only between five and ten deaths a year. There are 130 or so snakes in Australia, of which 60 odd are elapids, 24 are sea snakes, and about 6 are back-fanged colubrids. In Israel, in the

(top right) *The Cobalt Tarantula from Thailand is a vivid blue in some lights* (below right) *The European Scorpion looks far more dangerous than it is and need not be feared*

(page 86) *Man has an ambivalent attitude towards bees. While some bees provide honey, others can be dangerous. These are Honey Bees swarming*

years 1955-7 there were 418 reported cases of snakebite, of which only 17 resulted in the death of the victim.

FIRST AID

In Britain the only venomous snake is the Adder, and very few people have died as a result of Adder bite, in spite of the general fear of this unassuming little reptile. Even if someone should be lucky enough to come across an Adder, they are extremely unlikely to be bitten, but if this does happen it is as well to know what first-aid measures should be taken. The important thing is to reassure the victim that he is not going to die. The bite might well hurt, but it is not going to kill. This will prevent him from complicating the issue with needless anxiety and if you can calm him you will slow the heartbeat so that the venom will not be transported around the body as quickly as if he is in a panic. Forget all about tourniquets, slashing the bite open with a penknife, sucking the venom from the wound and all the other macho and harmful things we have been brought up to think of as correct. Simply wipe the site of the bite and cover it with a clean handkerchief. Provided you can get to a hospital within about half an hour that is all you need do, but keep the victim quiet and get a vehicle as close to him as possible rather than ask him to rush halfway across the common to the car. The activity will start the venom travelling through the bloodstream faster than it needs to. However, if it is likely to be quite a long time before you get to a hospital it might be as well to tie a firm, but not tight, ligature just above the site of the bite. There is no need to loosen it on the way, and anything you have handy can be used, such as a shoelace, or even a good twist of grass.

Until you get to a hospital, which you should always do as soon as you can, try to ensure that the bitten limb moves as little as possible. On the way if the patient vomits or retches, turn him on his side so that he cannot choke. When you get to the hospital, leave everything

(page 87: top) *In the tropics an encounter with a venomous animal can be fatal. Elsewhere it is usually only horribly painful; A nest of wood ants. These insects can give nasty bites but are not dangerous to man* (below) *Ralph Fitchett of the Tropical Butterfly Garden looking anew at Giant Millipedes on discovering that they produce a venom cocktail that includes cyanide*

(left) *Hornets are greatly feared, usually unnecessarily. Other insects mimic their colouration to avoid being eaten*

in the hands of the staff. The only other thing you can do is to telephone the hospital before you set out with the patient to tell them what has happened, but do not even think of doing this if it will slow you down at all. When someone you are with is bitten by an Adder it is very easy to become alarmed and convey your fear to the victim, though this can only do him more harm. It is particularly frightening if the person who has been bitten collapses, especially if you then discover that you cannot find a pulse. This early collapse usually resolves spontaneously within half an hour without any treatment.

In other parts of the world where far more dangerous species of snake are to be found, much the same sort of calm, thoughtful care for the victim while you transport him to hospital is the best thing you can do. In addition you can take a few measures to slow down the damage to his system. Again, you should not start cutting, sucking or tying. Don't wash any venom from the skin. It could be a handy way of identifying the animal that caused the bite. If you are able to kill the snake or other animal without any problem, do so in order that it can be identified and the appropriate antivenom applied, but only do this if there is no danger involved. If you have to hunt for it through the woodpile there is a distinct possibility that you too will end up a casualty, and that you will also waste a lot of time unnecessarily. There is no need to kill the animal if you know exactly what it is, but if you do kill it remember to take it with you. It has been known in this sort of situation that a car full of patient, friends and helpers have been halfway to hospital when somebody has remembered that the dead snake has been left behind and the whole circus has turned round to go back and collect it. Many people cease to think straight in an emergency.

When you discover that someone has been bitten or stung, you must immediately apply something that will act as a bandage around the injured limb. A crepe or elastic bandage is ideal, but failing that anything will do. The bandage should only be as tight as one would apply to a sprain. It has been found that if the affected limb is constricted slightly in this way, the absorption of the venom is slowed considerably, much more than by using a tourniquet, which can itself cause damage. As much of the limb should be bandaged as possible. When you have done all that, splint the affected limb, by using a piece of wood or something similar, or by tying one leg to the other, or an arm to the body – anything to immobilise it as far as possible. Recent research shows that this treatment allows very little venom to reach the bloodstream. It only remains now to get the patient to a hospital as soon as possible.

Everything that has been said about first aid and general protection against snakebite can be applied equally to most other venomous animals, though there are one or two important exceptions. In the event of a bite from a Red-backed Spider, the suggestions about immobilising the affected limb with crepe bandages should not be followed. It is best not to attempt any first aid but to get the patient to hospital as soon as possible. It has been discovered that the venom is very slow working and that if a bitten limb is restricted, local pain may become extremely acute.

Somebody suffering from the bite of a Blue-ringed Octopus or a sting from one of the cone shells should be treated as recommended for snakebite with the possible addition of the need for prolonged artificial respiration. Following the sting of one of the box jellyfish such as the Sea Wasp, vinegar should be poured over any adhering tentacles to inactivate them. Pressure bandages should not be applied nor immobilisation practised, and artificial respiration and cardiac massage may be necessary. Stings from venomous fish ought not to be treated by restricting the injured part in any way.

Most bites from snakes, or for that matter stings from other animals, are to one of the limbs. Treatment is a lot more difficult if someone is bitten on the body, or on the head and neck. If this has happened, do not try any first-aid measures, concentrate rather on getting the patient to hospital as soon as you can. Always remember that the patient is not likely to die, providing one gets him to a doctor.

PRECAUTIONS AGAINST SNAKEBITE

The possibility of snakebite is not really much of a problem if one is thoughtful and careful, and there are several precautions that everyone should take if they are likely to be travelling in an area of the tropics or sub-tropics where there are known to be dangerous snakes. Firstly, make sure you know where the nearest hospital is if you are going to be out in the field and away from civilisation for a while. If you are going to be really isolated for a long time it makes quite a lot of sense to take some sort of radio transmitter/receiver with you. Both these points are commonsense in this sort of situation anyway, regardless of the threat from snakes. Try and make sure that every member of your party is fit, and not overweight or suffering from a heart condition. When you get to where you are going, do not travel any distance on your own. It is far more sensible to travel with a companion. Always wear comfortable clothing and shoes, and if you are walking through undergrowth during the daytime, keep half an eye on where you are going to put your next foot all the time. Watch those people who live in such places and you will see that they do it constantly.

Perhaps the most sensible bit of advice is to leave snakes alone, and they will certainly reciprocate. If you have to lift stones and bits of bark, make sure that your fingers are not sticking over the edge, and that the stone is lifted away from you so that any snake that might be hiding beneath does not suddenly find a human face two feet away from it when its cover is raised. If you come to a log that you have to step over, first of all climb on top of it so that you can see what is on the other side. Do not walk about in undergrowth or long grass at night time or, if you must, do so with the aid of a powerful torch. Many snakes are active at night, and you cannot see as well as they can. If you live somewhere like this, make sure that any grass in your garden, especially around the house, is well cut and take all steps to keep your property clear of rats and mice, for they will attract snakes. If you have children, never let them collect or handle snakes until they are old enough and knowledgeable and experienced enough to know precisely what they are doing, and finally, if a young child comes and tells you that he has seen a snake, or touched a snake, or especially been bitten by a snake, please believe him. It might not be true, but he will not come and tell you when it is if you do not take him seriously and sympathetically every time.

FOLK REMEDIES

Even today when overwhelming evidence has shown that the only real treatment for severe cases of snakebite is a massive dose of antivenom, there still persists a host of folk remedies.

Some equally strange folk remedies were designed to immunise a person against the bite or sting of a particular animal rather than try to cure someone already bitten. In north Africa there is a theory that one becomes immune to snakebites after eating a live venomous snake, following which ceremony the initiate is able to summon snakes whenever he wishes. The calling of snakes is a reputed skill, and most snake charmers will insist that they have the ability, though the method of obtaining this miraculous power varies from place to place. The Bushmen of southern Africa drink the venom of snakes and chew the dried venom glands of the animals in order to render themselves immune to snakebite. They claim that it works and at the same time produces a sense of mild intoxication. The eating of the brains of snakes is supposed to give protection against the bites of venomous snakes, as is chewing tobacco. Since nicotine is a pretty powerful toxin it is more likely to do damage to the person who is taking it. Some say that the drinking of two tots of paraffin is a sure cure for rattlesnake bite. Followers of the rattlesnake cults of the south-eastern US, described in Chapter 10, firmly believe that their faith will save them in the event of being bitten, and over the years a variety of talismans and amulets have been thought to give protection against the bites of venomous snakes. Zulus go to considerable trouble to innoculate themselves against snakebite. Each year they perform lengthy ceremonies whose essential part is the mixing of venom with human saliva and smearing it onto a scratch on the back of the hand where a sliver of skin has been sliced off. Such a degree of immunity as this imparts is not permanent and the whole business has to be repeated each year.

Tales about the effects of snakebite, and degrees of venomosity, are legion. One of the pit vipers in Taiwan is known locally as the Hyappoda, or Hundred Pace Snake, a name which refers to the distance a man is supposed to be able to walk before dropping dead, and in central South America one is constantly told that the bite of a Jararacussu results in a fracture of the neck. This is not true but comes about because the victim's neck muscles relax and he no longer has any control over them, with the result that the head simply lolls about.

It is easy to understand how such a tale came to be believed, and most stories about venomous snakes are equally explicable even though many of them might be grossly exaggerated, but it always comes as a surprise to discover stories such as that about the Common Krait in Pakistan, where it is known as the Sangchul. There, they say, the snake does not kill by biting victims but by waiting until they are asleep and then sucking their breath until they are asphyxiated. In the villages where most of these stories originate, one finds that though the villagers are not zoologists they are superb naturalists, and one would imagine that they cannot fail to be aware of the fact that kraits bite, though rumours of asphyxiation makes some sense since we have already seen that elapids, of which the krait is one, cause death ultimately by stopping a victim from breathing.

Not far from Bombay, villagers will tell you that a Russell's Viper, should you come across one, will do you no harm provided you speak pleasantly and stop to pass the time of day for a while. Like most stories, this one is partly understandable. Snakes that are startled will bite as a reflex, but if one does surprise a Russell's Viper, and then stands quite still for a while, talking to it, the snake will slowly relax because no further threat is forthcoming, and will usually sneak slowly away to find somewhere quieter. The passerby can walk on, confident in the knowledge that the age-old advice works.

3 · The Trade in Venomous Animals

It is often difficult to realise the extent of the trade in venomous animals. But travel to other parts of the world and you soon discover the extent of what goes on. The trade can be divided in the first place into live and dead stock. The live animal trade is very small indeed, with two exceptions, because not many individuals or zoos around the world require large numbers of venomous beasts. The two exceptions that are traded in any quantity are snakes, and to a lesser extent animals like scorpions and spiders which trappers supply to snake farms. The trade in cobras from places like Thailand to Singapore, Taiwan, Hong Kong and other cities with large Chinese populations is astonishingly active, and takes place because of the Chinese belief that various parts of venomous snakes are efficacious in the treatment of a variety of diseases.

The trade in dead venomous animals is colossal. As with live cobras, Chinese pharmacies buy considerable quantities of dead material with which to stock their shelves. For some reason, vipers and pit vipers do not feature to any extent in this trade, though the odd Russell's Viper does. Most of the trade is still in cobras, together with scorpions and centipedes which most pharmacists seem to stock by the drawerful. Throughout the world this trade must account for many tens of thousands of animals a year, but the larger threat is the demand for the skins of venomous snakes for the fashion and curio trade. The Far East is without doubt the centre of this business. It is true that there is considerable trade from many parts of the world in the skins of reptiles of various sorts, but they are mainly non-venomous snakes such as pythons, boas, rat snakes and so on, as well as crocodilians and monitor lizards and, funnily enough, amphibians.

It is a rare hotel in India, Thailand, Malaysia, Singapore or Indonesia in which the skins of venomous reptiles are not for sale, either as skins, or usually in the form of made-up products. The most common products are ladies' handbags, briefcases, wallets, men's and women's purses, and

SIAM FARM & ZOOLOGICAL CO., LTD.
1875/145 Charansanitwong Road,
Soi Lert Boon, Bangplad,
Bangkok (Zip Code) 10700
Thailand.

Telephone: 424-6129
Telex: 84968 SAMFARM TH

Bankers: Bangkok Bank (Rajvithi Branch) Acc. No. 131302955-

POISONOUS SNAKE   PRICE LIST   EACH (US$)

Common name	Scientific name	Each (US$)
Blue Kraits	Bungarus candidus	US$ 20.–
Banded Kraits	Bungarus fasciatus	7.–
Red headed Kraits	Bungarus flaviceps	60.–
Elegant Coral Snakes	Calliophis maculiceps	15.–
Malayan Pit Vipers	Calloselasma rhodostoma	5.–
Blue Long-glanded Coral Snakes	Maticora bivirgata flaviceps	15.–
Black and White Spitting Cobra	Naja n. atra	8.–
Black Spitting Cobras	Naja n. atra (Var)	8.–
Issan Spitting Cobras	Naja n. isanensis	10.–
Monocellate Cobras	Naja n. kaouthia	8.–
Suphan Cobras	Naja n. kaouthia, variety B	40.–
Golden Spitting Cobras	Naja n. sputatrix	20.–
King Cobras, under 6 feet	Ophiophagus hannah	60.–
over 6 feet	" "	80.–
Kanchanaburi Banded King Cobra	" "	
under 6 feet	" "	80.–
over 6 feet	" "	100.–
Siamese Russell's vipers	Vipera russellii siamensis	7.–

MILDLY POISONOUS SNAKES

Common name	Scientific name	Each (US$)
Green Cat-eyed Snakes	Boiga cyanea	3.5
Dog-toothed Cat Snakes	Boiga cynodon	5.–
Gray Dog-toothed Cat Snakes	Boiga cynodon siamensis	10.–
Black Dog-toothed Cat Snakes	Boiga cynodon (Var.)	6.–
Mangrove Snakes	Boiga dendrophila	6.–
Jungle Whip Cat Snakes	Boiga drapiezii	5.–
Yellow-chined Cat Snakes	Boiga jaspedea	5.–
Marble Cat Snakes	Boiga multomaculata	7.–
Red Cat-eyed Snakes	Boiga nigriceps	12.–
Mock False Vipers	Psammodynastes pulverulentus	5.–
Green Pit Vipers	Trimeresurus albolabris	3.–
White lined Pit Vipers	Trimeresurus erythrurus	10.–
Kanburee Pit Vipers	Trimeresurus kanburiensis	10.–
Mountain Pit Vipers	Trimeresurus monticola (Rare)	150.–
Popes's Pit Vipers	Trimeresurus popeorum	4.–
Brown Flat-nosed Pit Vipers	Trimeresurus puniceus	45.–
Mangrove Pit Vipers	Trimeresurus purpureomaculatus	8.–
Bamboo Pit Vipers	Trimeresurus stejnegeri	6.–

Animal dealers around the world sell venomous animals for the preparation of antivenoms (John Nichol)

A heap of serpents awaiting sale in a snake shop in South Korea (Herbert Lieske)

passport and credit card holders. Items of clothing come second with boots the most popular item of all, followed by belts, shoes and, would you believe, trousers, waistcoats and jackets. If you had hitherto believed that the stage costumes of Dame Edna Everage represented the epitome of kitsch, you have never seen a dyed cobra-skin, full length jacket. One would not think that anyone today would be insensitive enough to wear reptile skins but it is surprising how great a trade there is in them. In Bangkok alone there are hundreds of shops selling cobra skin, and some Russell's Viper skin products, and most of the shops along either side of Sukhumvit Road stock these items.

Following bags and clothes, the next group of products made from snakeskins must be called generically, souvenirs. The commonest of these are stuffed cobras, or stuffed cobra/mongoose pairs, and all the shops that sell belts and wallets sell these as well. For the comparatively impecunious collector there are keyrings and other bits and pieces made from the scraps of skin left from the manufacture of other, larger items. The factories which manufacture all these things are awful places. There are several around Samutprakan in Thailand, where the animals are skinned and the skins are processed. The skins normally go elsewhere for making up. Quite a number of the skins that these places receive are a by-product from the suppliers of snake meat and other medicinal cure-alls.

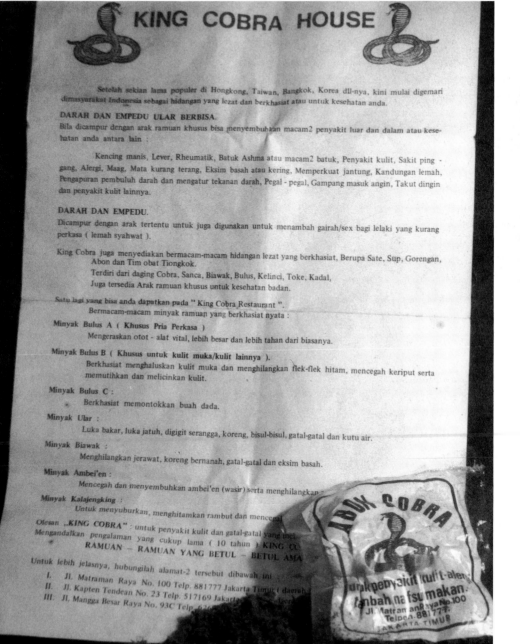

In Indonesia there are restaurants that sell dishes from cobra meat, or one can buy a packet of minced cobra to take home (John Nichol)

One might have thought that Russell's Viper would have featured prominently in the skin trade since it has a highly patterned and most attractive skin. But again, most of the reptiles used are cobras, due no doubt to their reputation, and to some extent to the pattern on the hood, though Thai and other cobras from the eastern part of their range do not have the distinctive 'spectacles' mark that most Indian Cobras possess. In Thailand, cobras are brought into the processing plants in large numbers. They are suspended by the neck and skinned alive. Just below the head the spinal column is severed and the body is thrown into a bin. It is unlikely that these bodies are wasted, presumably they are either sold to Chinese pharmacies, or as animal food. Some are certainly desiccated and sold for medicinal purposes. The tanned skins are sold to factories where rows of poorly paid workers spend virtually all day between sleeping and eating, turning out souvenirs for the tourists.

In 1986 it became illegal in India to trade in cobras and Russell's Vipers. Henceforth the only way in which a customer will be able to obtain the skin of this sort of animal will be to go directly to the official Indian body, the Bharat Leather Corporation, which will have full control of the trade. Until now the skins of many snakes were exported annually, and products made from them were freely available in all the tourist centres. India, however, has legislation forbidding trade in many other wild animal products which are still available if you know where to look, so it will be interesting to see what happens over this bit of new legislation. There is no demand within the country for snakeskin products, and though one never actually sees foreign tourists buying this sort of thing in India, there must be a healthy trade as the shops selling it are legion.

For some time there has been a persistent illegal trade in the export of snakeskins. There is no way of assessing the extent of the business, Customs' figures fluctuate wildly from one year to another due to variations in the snake population and the state of the local economy In India in 1980, 166,653 skins were intercepted, in 1981 the figure was 183,950, in 1982 it was 74,189. In 1983 a total of over 204,000 skins were stopped, in 1984 the figure was about a million, and in 1985 (the last year for which figures are available) 72,604 skins were seized at Cochin, Delhi and Bombay. By no means all of these were of venomous snakes, in fact only a small quantity were, but the figures are interesting for all that. There can be no doubt that the illegal trade will continue, for Customs officers do not consider it nearly as important as the smuggling of gold and narcotics, and most of the seizures of snakeskins were the results of 'lucky accidents'.

It would appear that most illegal consignments from India go to either

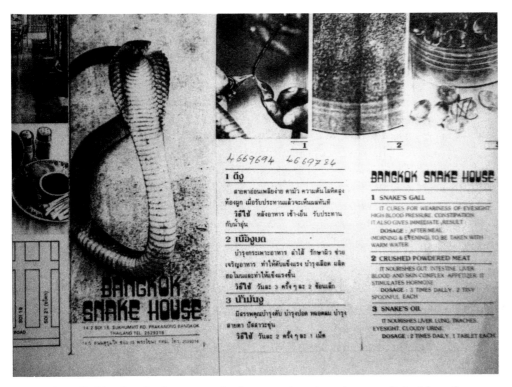

The gall, blood and other parts of cobras are thought to have medicinal benefits in some parts of the world (John Nichol)

Singapore or the United Arab Emirates, whence they are shipped to other parts of the world. This raises difficult questions when they reach their destination since it is well known that the skins do not originate in these countries even if the documents accompanying the consignments declare that they do. There is a well-established smuggling trade route between the west coast of India and the Middle East. Small dhows smuggle a range of commodities, especially gold and electrical goods, from the Middle East to India and carry drugs and snakeskins on the way back, goods that ultimately find their way into Western countries. The business is highly organised, and it has been estimated that the annual value of snakeskins entering the Middle East from India is something like £60 million. There is also considerable trade in snakeskins between India and Bangladesh and, interestingly enough, the only snake skins that one can legally export from Bangladesh are those of venomous species.

Other venomous animals are traded as dead specimens in various parts of the world for tourist souvenirs, and though quite a number are used for this purpose, compared with the trade in snakeskins the number is small.

In most parts of the tropics and sub-tropics, wherever one finds tourists from Europe and the USA, there are shops selling dried specimens of local scorpions and spiders. Mounted scorpions of more than one species may be bought throughout north Africa, and the larger scorpions such as the Emperor Scorpion can be found all over the world, mounted in small display cabinets, very often miles from their actual range. There is a similar trade in spiders, particularly those of the family Theraphosidae, the large, spectacular bird-eating spiders. One particular collector in South America catches large numbers of *Theraphosa blondi* on the marshes near his home in French Guiana. He keeps them, dead, in large drums full of alcohol until he is ready to set them. In some places both scorpions and spiders are increasingly sold embedded in clear plastic.

In Thailand it is very easy to buy Long-nosed Green Whip Snakes in the animal market at Bangkok for 5 baht each (15p or 7.5 cents). The losses must be horrific since they are not the easiest snakes to keep. They are delightful, elegant things but they require specific conditions and will usually only feed on lizards, frequently just geckos. Since in that market they are sold mostly for passing-impulse trade, virtually all of them must be dead within two or three weeks.

It is in Thailand and in Hong Kong that one can most easily watch the preparation of cobra products. According to promotional literature from the Bangkok Snake House, the following parts of the anatomy of a cobra are used for medicinal purposes:

Snake's Gall: It cures for weariness of eyesight, high blood pressure, constipation. It also gives immediate result. Dosage: After meal (morning and evening) to be taken with warm water.

Crushed, Powdered Meat: It nourishes gut, intestine, liver, blood and skin complex. Appetizer. It stimulate hormone. Dosage: 3 times daily, 2 tiny spoonful each.

Snake's Oil: It nourishes liver, lung, trachea, eyesight, cloudy urine. Dosage: 2 times daily, 1 tablet each [sic].

The whole business is gruesome in the extreme. The snake is hung by the neck and the tail is pulled out straight so that the animal cannot wriggle. An incision is made from throat to cloaca, and a major blood vessel carefully extracted with the aid of a pair of forceps. A small nick is made into the vein and blood is allowed to drip from it into a wine glass. To this is added the contents of the gall bladder, and the mixture stirred with a slug of brandy. The still squirming body is cut away, leaving the head and the skin hanging from the noose. The body is either added

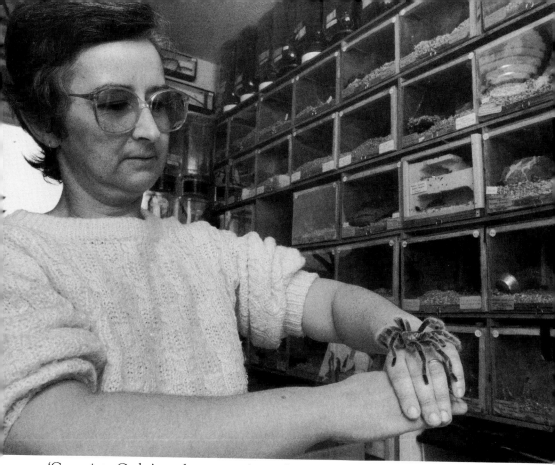

'Come into Cathy's parlour . . . if you dare', was the newspaper headline about a woman breeding invertebrates commercially (John McLellan/Yellow Advertiser)

to the stock pot or, if the customer requests it, given to him in a smart carrier bag. The skin, with the head attached, is salted and sent off to a tannery to be made into someone's handbag.

One of the most bizarre ornaments produced from the skin of a cobra was spotted in Jakarta, where it had been stuffed in the normal, erect position. In this instance however, the eyes had been replaced with tiny red lamps. The lead was passed down through the body to emerge at the cloaca. When it was connected properly to a disco system the eyes flashed in synchronisation with the music. Only in Jakarta do they appear to use Mangrove Snakes for medicinal, or at any rate culinary, reasons in addition to the usual cobras, Russell's Vipers and the occasional python.

Many of the establishments that maintain stocks of snakes for venom production make a bit of extra money by allowing the paying public to watch some of the snakes' activities. Such exhibitions always seem to attract substantial audiences, due no doubt to the creepy fascination

Puffer fish inflate themselves to discourage threats. Their poisonous flesh is considered a delicacy in Japan where it is known as Fugu (Lee Ironside)

these animals hold for people. Many of these places are in tropical countries where the gullible population is fed a running commentary in the prose that used to be found in Victorian melodramas, and much of the information put across is pure rubbish. At one such place in Thailand the visitors are told that kraits at night will chase humans and bite them just as a dog does. As we have seen, a krait at night can be a dangerous customer, but that sort of statement is absurd. A newspaper report on this same place states that 'The snakes slide and dart across the ground, flicking forked tongues at the crowd, to the horror of most adults and screams of delight from the children.'

Another trade in venomous animals is that in corals and the shells of sea urchins for the jewellery and tourist trades. Each year hundreds of tons of dried corals are sold, often illegally, to end up as souvenirs on someone's mantelpiece or in aquaria. In the Philippines, parts of Asia, Brazil and the Maldives, corals are collected and used in road construction, and in the production of lime, calcium carbide and cement. Most parts of the world where corals are to be found, export them, even though they may be protected, and there is evidence that such depletion of stocks is beginning to cause concern. The trade in sea urchins appears to be solely in the form of ornamental items as tourist souvenirs. Almost always the urchin shell is emptied of all tissue, and the spines are removed, leaving the attractive ornamental globe. Sometimes these are made up into gimmicky lampshades. Sea urchins, incidentally, are collected as food in some areas.

Finally, as was mentioned in the Introduction, some venomous fish are sold as food in markets round the world, but the quantities involved are tiny, with the exception of the various puffer fish purchased in Far Eastern restaurants as *fugu*. Far more puffer fish are caught than any other species, but even so the trade is no threat to the animals as a whole. Dried, inflated puffer fish are also available as souvenirs for tourists.

4 · Venomous Animals
and Religion

Throughout history animals have played a part in the worship of gods of many descriptions, and even where they have not had any especial religious significance, they are still sometimes used as sacrifices in ceremonies. The use of specifically venomous animals is surprisingly restricted, given man's fascination with them. With the exception of the Marine Toad, most venomous animals with any sort of religious significance seem to be snakes. In west Africa, pythons are revered, but apart from this one exception, all other snakes with a sacred reputation seem to be venomous.

On the island of Haiti there is a long history of voodoo – or voudoun, which is the current, trendy way of spelling it. If you have been watching all the right horror movies you will be aware of the story of zombies that is associated with voodooism. According to legend a zombie is a person who has died, and by the power of the shaman conferred on him by Baron Samedi, is made to rise from the tomb to become one of the 'living dead', destined to spend his time fulfilling the wishes of the priest. For many years this tale had been taken with a pinch of salt by other nations despite its persistence, until some years ago researchers began to take the story seriously enough to wonder if there was anything in it. In due course it emerged that what generally happened was that someone, very often a member of the zombie's family (and sometimes all of them together), would go to a shaman with a grievance against his relative and ask for him - or her for that matter - to be made into a zombie. This decision was never taken lightly or frivolously, but only after considerable family discussion and all other options had failed. It was generally when the person concerned had committed so heinous an offence against the family that they felt nothing else would do, and was usually the result of years of family discord and tension.

(right) *The caterpillars of the Cinnabar Moth are poisonous but not venomous. Their vivid colouration warns predators*

The venom from South American Poison Arrow Frogs is used to tip the arrows of hunters

According to the research, the shaman for a fee would prepare a compound that would be fed or otherwise introduced into a victim which caused him to appear to die. All this was accompanied by religious ceremonies, dancing and the sacrificing of a live white cockerel - a bit of set dressing never did any religious ceremony any harm. Finally, the apparently dead man would be buried. In due course he would come back to life and dig himself out of his shallow grave. This sounds pretty impossible, but it seems that when a person was to be made into a zombie he was often buried under only a few centimetres of dust, perhaps with his head above the surface. Evidence showed that the victim would then resume life, wandering about in a dozy state, ready to obey the commands of either the shaman or the person who had requested the whole thing. It soon became evident that the zombie had not really died at all, but had been in a catatonic state, induced by the preparation given to him by the shaman. Eventually he would slowly recover and would once again be normal, though he would often have to move to another area to live because his family would no longer accept him and other members of the community were uneasy at living in the same village as a zombie.

One particular researcher decided to look into the ingredients of the magical medicine that was necessary to turn someone into a zombie, and after a few blind alleys and wrong guesses, found that it consisted of a load of rubbish that served no purpose at all, ingredients included for effect or simply as fillers. Originally he tried to discover whether there was any connection between it and some of the ju-ju recipes from west Africa, whence the religion originated, and thought that the Calabar Bean and Datura, both poisonous plants, might be responsible for the state of the zombies. He could find no Calabar Bean growing in Haiti. He did, however, discover Datura, and found that this material was used in the shaman's recipe. What finally emerged at the end of years of research was that the most virulent ingredient of the mixture was not Datura after all, but the venom of the Marine Toad. The Haitians call it the Bouga Toad. It is a common amphibian throughout its range, and has large paratoid glands that produce the venom. A few live toads were thrown into the stew while it was cooking, and the toxic ingredients from them resulted in the apparent death of the victim and his subsequent catatonia.

Toad venom is often underrated, but as long ago as Roman times, women were using toads to poison their husbands. The species used was probably the European Toad, and all over Europe from that time onward there was considerable interest in developing toad venoms as a poison to get rid of one's enemies. The method of extracting the venom was originally to throw a whole heap of live toads into a pan of boiling

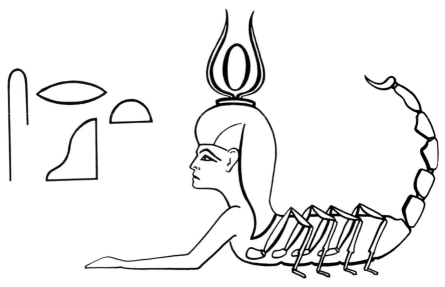

Selkis, the scorpion goddess of ancient Egypt (Michael Beasley)

water. The venoms would rise to the top and could be skimmed off, and in the sixteenth century Italian poisoners were busy perfecting a method of incorporating toad venom into salt that could be sprinkled onto a victim's food without detection. By the beginning of the eighteenth century toad venoms were considered such an efficient way of killing people that it was added to the contents of explosive artillery shells.

At the same time as all this was going on, 'doctors' were making use of toad venom to treat a whole spectrum of ills from toothache to sinusitis to the common cold, but as ever it was the early Chinese physicians who had perfected the preparation for medical usage. They would boil it until it became a thick paste and then form it into smooth discs which they called *ch' anu su* – toad venom. Its virulence is due to a variety of ingredients. One of them is a compound known to the modern chemist as bufotenin, a hallucinogenic drug.

When one researches into the Mayan and other Indian cultures of central America, two things become apparent. One is that much of their religious art depicts large toads which are almost without doubt Marine Toads, and the second is that when the middens of Mayan habitations are excavated, great heaps of Marine Toad remains can be found. This has led to a theory that these people used the venom of Marine Toads as a narcotic in their religious ceremonies. Some scientists refuse to believe this, and suggest that the Mayans simply ate the toads after cutting away the venom glands, as is done today by the Campa tribe of Indians from the Upper Amazon. Whatever the case, other Indians from that part of the world partake of a hallucinogenic snuff made from a plant which contains exactly the same compound – bufotenin.

The ancient Egyptians used venomous animals as religious symbols in a big way. According to the religious teachings of that venerable culture, the creator of all things was Ra. After a long life Ra was growing old and one of his many grandchildren, the goddess Isis, longed for the power that only he had over the world; no one but Ra was able to create anything. One day Isis collected a small handful of soil on which had fallen some saliva from the mouth of Ra. She modelled the wet soil into the form of a snake, and spent much time speaking to it and whispering spells. Eventually she placed the serpent where she knew Ra would walk, and when he passed the snake it came to life and bit him in the ankle, and once more crumbled into dust. The bite was venomous and Ra began to suffer. In the words of the Papyrus of Ebers, written about 1550BC, Ra cried:

I have been stung by a serpent which I could not see. This is not the same as fire; it is not the same as water, but still I am as cold as water and then again I am as hot as fire. All my body sweats, and I tremble. My eyesight is not steady and I cannot see for the sweat pours over my face in spite of summer's pleasant air.

This piece of writing cannot be bettered as a description of the bite of a snake with neurotoxic venom, and is thought to refer to an Egyptian Cobra.

Ra summoned all his sons and daughters to help him and they came from all corners of the world, including Wadjet the cobra goddess, Selkis the scorpion goddess and the snake-headed goddess of the harvest, Renenutet. In the end, Isis herself chanted a spell to drive out the venom so that Ra recovered completely, and Isis had gained much in his esteem. When Ra was well again after his suffering he determined that in future man would not have to put up with this sort of pain at the expense of venomous serpents and he made a treaty with them and they agreed not to bite man thereafter. But in due course the snakes reneged on this treaty, so instead Ra gave to man the magic word *hekau*, which when spoken would ward off any venomous snakes. The people who were the first to receive this word from Ra were the Psylli, a people of north Africa. They became the first known snake charmers, and were much in demand to clear houses of dangerous snakes, but it was said of them that they smelled bad – a charge that is frequently levelled at snake charmers today. Some denigrators may say that their smell alone will kill snakes!

The pharoahs of Ancient Eygpt incorporated a vulture and a cobra in their headdresses (John Nichol)

In passing, it is of interest that whenever arguments arose regarding the paternity of a baby among the Psylli, (whether a baby was fathered by one of them or by an outsider), they would expose the poor little mite to venomous snakes to test the claim. If the snakes did not bite it was deemed to be pure Psylli.

In ancient Egypt the valley of the Nile was prone to flooding, and whenever this happening coincided with spring and the melting of snows in the mountains far to the south, cobras, and no doubt other snakes, were washed from their holes. This seasonal appearance of hordes of cobras came to be associated in the minds of the Egyptians with a time of growth, and thus the cobra became the symbol of fertility. As a result these snakes were considered of great importance in their religious rites. Cobra jewellery was worn as a charm, and the four corners of the paradise of the ancient Egyptians were said to be guarded by fire-breathing cobras, while Horus the hawk-headed god is depicted as holding cobras, and was known as the Stopper of Snakes.

The already mentioned goddess daughter of Ra, called Wadjet, is a fire-spitting cobra, sometimes known as Uraeus, which means 'she who rears up'. She was the national goddess of Lower Egypt, whose counter-part in Upper Egypt was Nekhbet the vulture (who could also assume the form of a serpent). The two of them appear on the headdresses of the pharaohs representing the unity of Egypt. Wadjet was also supposed to be able, whilst she was on the pharaoh's forehead, to spit a sort of venomous laser beam at any of his enemies who might approach with evil intent. The ruler was always known as The Lord Of Two Lands, The King of Upper And Lower Egypt. Trying to decipher Egyptian religious mythology is no easy matter, but it would appear that all the snakes referred to therein are cobras. Most are likely to be the Egyptian Cobras but some could be Spitting Cobras.

There has always been a close relationship between snakes and Egypt. The Greek writer Diodorus Siculus claimed that snakes were generated spontaneously by the shifting sands of Egypt, though it was more likely that he was referring to vipers of some sort rather than cobras. Many viperids are extremely well camouflaged against sandy backgrounds, and some of them shuffle themselves down into the sand to hide. Ancient Egyptians believed that someone bitten by a venomous snake was there-after profaned. Snakebite was regarded so seriously that, in the tombs, mummies were provided with amulets to protect them against snakebite in the afterworld. It was also thought that if a snake did bite a man, it was forever shunned by all. Even the earth refused it sanctuary, and it could never return to its hole. To this day the dread of snakebite is such

Religions the world over involve venomous animals. This Hindu god, Ganesh, is protected by a multi-headed cobra (John Nichol)

that contemporary Egyptians will draw a circle round their bed before retiring for the night. Sometimes an engraved wand is reserved for this rite, which protects the sleeper from snakes and scorpions until morning. A fascinating similarity of behaviour can be found among the cowboys of America who until recently, and perhaps even to this day, would coil their lassoo in a circle round their sleeping blanket at night before settling down, in order to protect them from the bites of rattlesnakes.

Just as the religion of ancient Egypt seems to be obsessed with cobras, so too do some aspects of Hinduism. Hindu women leave gifts of milk, yoghurt, saffron and honey by cobra holes as an offering to the snake and as a prayer for fertility, and in Bombay, Mysore and the Kathiawar Peninsula, a childless woman will hang a cobra carved from stone, known as a *nagakal,* inside a well for a period of six months. If the woman conceives during that time, she builds a shrine for the carving beneath a sacred pipal tree. There is a secondary benefit to this ritual as it is said that the soil from around such a shrine is beneficial in the treatment of leprosy. Religion in India involves every aspect of life, and everywhere one can find small religious models of a phallus which is worshipped as the emblem of the fertility god, Shiva. On occasion the phallus is to be found in the form of an erect cobra.

The people of central India believe that cobras are descended from the Nagas, who were the serpent gods of Bharat, the local name for India. Every garden in Nagpur and along the Malabar coast has a wild corner set aside for cobras, who are encouraged to take up residence. Such an area is known as a *vishahum kavu,* a poison shrine, and food and other offerings are put out for the cobras, who are worshipped regularly. The state of Rajasthan is famous throughout India for its snake charmers, and in the Punjab the month of September is given over to snake worship, while 20 July is the feast of Nag Panchami, a day when the devout release cobras to gain merit in the next world. The Indian words *nag* and *naga*, and the female, *nagini*, are synonymous with *Naja*, the generic name of the cobra. It is said that the rajahs of the town of Nagpur are descended from a cobra.

Buddhists, too incorporate cobras into part of their religious teachings, and tell a story about Muchilinda, the cobra, who went to the aid of the Lord Buddha when he was suffering from heatstroke. Muchilinda spread his hood over Buddha to protect him from the sun, and by way of thanks Buddha reached out his hand and touched the cobra in blessing, leaving the marks of his fingers on the hood, which are there to this day. Buddha obviously had a good relationship with cobras because on another occasion he came to the bank of the sacred River Ganges and wished to cross, but along its length the river was too wild and too deep. A cobra heard him muttering to himself about being unable to get to the other side, and the snake called upon all other cobras to come to Lord Buddha's aid. When enough of them had arrived, they formed themselves into a living bridge so that Buddha could walk across their bodies to reach the other side. However, the cobras were so keen to help that more and more of them came pouring in from all directions until there were enough of them to form four bridges. Buddha, not wishing to disappoint any of the waiting snakes, suggested to them that they go ahead and form four bridges, whereupon he courteously divided himself into four separate Buddhas for the occasion, so that he could use all four bridges simultaneously. Once he had crossed the river he made his way to the holy town of Benares where he sat down and discussed the law with the great Naga Elapattra, a five-headed cobra, so large that his shadow covered the entire deer park at Benares.

In Pakistan there lives a tribe known as the Jogi, whose members are renowned throughout the country as snake charmers. Each of them, when he is learning his art, swears an oath to Gogol Vir the patron god of snakes. Two thousand miles to the east in Malaysia the Wagler's Pit Vipers are considered sacred, and as such are kept in temples where they are said to be fed on eggs. Local worshippers insist that this snake is too lazy to go out and find food for itself, so a little bird takes on the job of feeding it. In Japan it is believed that volcanoes are guarded by some species of supernatural venomous serpent, and in the Philippines the people of the Ifuyao tribe, which until fairly recently were head-hunters, incorporate into their religious beliefs a legend that cobras are weresnakes which behave in much the same manner as the better known werewolf.

In Cambodia, or Kampuchea as it is currently known, it is said that the Khmer kings are descended from a Naga princess, and long, picturesque tales about how this happened are told in this beautiful country that has been devastated in the last few years, leaving a nation of gentle people suffering untold horrors.

The Christian Bible is full of references to venomous animals, one in the Book of Kings. It has already been seen that the Ancient Egyptians considered cobras to be sacred, and the priests of the pharaohs took many decisions in relation to what was revealed by snakes. One of the reasons that the pharaoh of the time was displeased was that Moses, (and Aaron) were both highly respected snake shamans, and in some instances they had competed with Pharaoh's priests in this respect, and were besting them too often for comfort. Moses' trade mark was a bronze serpent on a pole. When Hezekiah became king of Israel he immediately started to reform society left, right and centre, and one of the things he did was to break up Moses' snake, which was called Nehustan. Until then, we are told, the people of Israel had burnt incense in honour of it.

Even the serpent in the Garden of Eden would appear to have been venomous. The story goes that the Devil wished to enter the garden in order to seduce Eve, but various angels were acting as security guards and despite considerable effort he was unable to find a way in. In the end he approached the snake and asked if he could hide inside one of his hollow teeth. The snake agreed in principle but wanted to know what was in it for him, and after a moment's thought the Devil promised that as a reward the snake would be given the sweetest food in the world. To this it agreed and the Devil squeezed into the hollow tooth, and hidden like this entered the garden.

We all know what happened next; the Devil, still hidden in the tooth of the serpent, asked Eve to taste the apple, which of course she did. When Adam turned up to make trouble for the Devil, the snake was told that he could not have his reward, which was human blood, unless he bit Adam, but the snake wanted proof that human blood actually was the sweetest food in the world. He refused to accept the Devil's word for it, and asked his friends, the mosquitoes, to go off and sample as many types of blood as they could, and to come back and let him know the results of their survey. In a while they returned, and the snake asked for their verdict. One of the mosquitoes was inclined to agree with the Devil that indeed the blood of man was the sweetest food, but just as he was about to say so, a swallow, divining what he was about to reveal, zipped down and bit his tongue, leaving it forked like a swallow's tail or a snake's tongue, depending on which version of the story you read. Therefore, when the mosquito tried to say what he thought, the word was distorted and came out as 'frog' which the snake assumed was therefore the best food in the world. And from that day snakes have preferred frogs to any other food. Before the serpent could wander off to find himself a frog, however, the archangel Michael descended in a fury, and in great anger chased away the Devil, who was still hidden in the tooth of the serpent, until both were out of the garden.

Snakes crop up in religion everywhere, even on the little island of Fernando Po off the west coast of Africa, where a cobra deity is worshipped as the guardian of the community. This snake god can bring blessings of all sorts to those who believe in it, but it is also responsible for retribution. In an annual ceremony a cobra skin is hung publicly in each village, and new babies are taken to touch its tail to bring them a fortunate life. The Ancient Greeks, too, incorporated snakes into their

The Kerykeion, the staff of the
messenger of the gods, Mercury,
depicts two stylised Adders
performing their so-called dance
(Michael Beasley)

religion, and one was a serpent familiar of the goddess Athene. It lived at her shrine in the Acropolis and was regarded as the guardian of the city. On one occasion Athene incited Perseus to kill Medusa – a particularly difficult task since merely glancing at her face was enough to cause instant death. Perseus was not to be outdone and, rather than look directly at Medusa, he used his polished shield as a mirror. Apparently this worked, as he managed to kill her by chopping off her head. He picked up this grisly souvenir and climbed onto his horse to fly to Africa, and as he did so, wherever a drop of blood from the head fell upon the earth it turned into a viper.

Another nice little story from Greece concerns Hygeia, the daughter of Asclepios the surgeon. Hygeia kept venomous snakes in her temple and fed them on milk. As a thank you for this thoughtful treatment the snakes used to restore the sight of blind patients by gently flicking them with their tongues. Hygeia's father used a non-venomous snake, nowadays called an Aescalupian Snake after him, as his symbol and this, curled round a staff, is used to this day as a sign of the medical profession. As such it is the insignia on the badge of the Royal Army Medical Corps. Another Greek, Hermes, messenger of the gods, had a staff of office in the form of two serpents wrapped round a central pole surmounted with a pair of wings. This token is sometimes known as a caduceus (Latin) though it is properly called a kerykeion (Greek). The form of this emblem is especially interesting since the snakes themselves are almost certainly stylised representations of Adders. In spring when Adders first emerge after their winter hibernation, males sometimes fight other males. At times they rear up, and twist the foreparts of their bodies about each other. They will wrestle like this for a while, trying to push their opponent to the ground. No harm is done to either animal, but the two entwined Adders have become the snakes on the kerykeion.

In modern Greece, peasants pour milk through a small hole in the floor of their houses as a libation to snakes. This hole is known as the snake's door, and in Lithuania snakes are fed milk and sprinkled with beer during the fertility rite.

If you are ever in the village of Cocullo in the Abruzzi mountains of Italy on 4 August, you will have the opportunity to watch a magnificent religious snake festival, for on that date in the year 1218, according to legend, Saint Dominic stopped in Cocullo, and in return for hospitality rid the village of a plague of vipers. Nowadays in the weeks before the festival to commemorate that event, the local population goes out into the surrounding countryside and collects as many snakes as it can. Both venomous and non-venomous species are caught, though it is only fair to say that the vast majority are completely harmless. They are carried in procession with the saint's statue through the town, and afterwards they are usually sold to animal dealers, tourists, and anyone with enough lira in his pockets. It is known that from the seventh century the populace of Lombardy worshipped a golden viper, and to this day the arms of the Visconti family of that region of Italy depicts a snake with a child in its mouth.

A saint whose relationship with snakes is better known than Saint Dominic's is Saint Patrick who, despite the fact that there were never any snakes there, has been credited with ridding Ireland of serpents. The Irish, being fiercely religious people, refuse to listen to scientific reasons for the absence of snakes, and believe implicitly in the Saint Patrick story. Therefore it was a very brave, or perhaps a very foolish, gentleman by the name of James Cleland who in the year 1831 bought half a dozen non-venomous snakes of an undisclosed species, though it is likely that they were Grass Snakes, from Covent Garden market in London and took them home to Ireland to release on his estate at Rath Gael in County Down. He wisely told no one what he had done, but some time later one was found, having wandered from the site of its release, and because no one would believe that there was such a thing in the country it was said to be a worm. Subsequently a further three turned up here and there, and one was taken to a knowledgeable person for identification. He told the finder that it was without doubt a snake, and when news of the discovery got out all hell broke loose. The local priest promised hell fire and damnation on anybody who neglected to kill any snakes they might come across, and a savage, religious fury took over for a while during which anything remotely resembling a snake was hacked to pieces amid much rejoicing. Most sensibly James Cleland kept quiet about his escapade.

Despite the fact that there are no snakes in Ireland, a large Bronze Age burial site at Newgrange has pictures showing coiled snakes. The Irish have also got the reputation for being able to immobilise Adders by drawing a ring round them, and it is also said that if one washes the site of the bite of an Adder with the milk of an Irish cow, it will cause no further problems to the victim.

Elsewhere in Europe religious mythology is rich in tales of venomous snakes, which are almost always Adders. Baltic legends are particularly abundant in snake stories, and the way they visualise hell is as a dark, icy chamber with walls that consist entirely of giant Adders whose heads face constantly inwards, and whose fangs forever drip venom to form a river in which the damned must swim for eternity. Norse mythology tells us that the home of the gods is at the top of a huge tree, and is known as Asgard. But while the gods go about their business there is always a mass of Adders that spend their time eating the roots.

The Indians of North America had respect for all living things with which they shared the world, and all had a special relationship with snakes. Rattlesnakes were particularly venerated, and played a large part in their lives. If an Indian ever killed one he had to ask for pardon from it or others would come to avenge the death upon him. Women in labour were given powdered snake rattles in a drink to make for easy delivery, and in some cases the skin of a rattlesnake was wrapped round the woman's abdomen for the same purpose. Most older religions incorporated medicines and healing processes, and all these American Indian remedies were used in conjunction with prayers and invocations from the medicine man who doubled as priest. The fangs of rattlesnakes were ground and used in the treatment of various illnesses, and they were sprinkled on the earth during the green corn dance of the Seminoles.

The Pomo Indians of California used to make a magic potion from the blood of four rattlesnakes and mixed it with spiders squashed to a pulp, adding similarly pulped bees, scorpions and ants. The whole dreadful mixture was dripped onto the effigy of an enemy, or used to tip an arrow that was fired over his hut, and the result was always death. However, if you didn't want to kill an enemy but were content with merely blinding him, this could be achieved by removing the eyes from a rattlesnake that was about to slough. The eyes were rubbed on the smooth internal surface of an abalone shell, and flashed onto the eyes of the intended victim with the desired result.

The Algonquin Indians have a story about the mighty Manitoa-ah, who was hunting one day and became very annoyed with his penis because it got in his way. In his anger he pulled it off and threw it away. When it landed on the ground it immediately turned into a rattlesnake and thereafter these reptiles were regarded by the Algonquin as symbols of life.

Religious beliefs and ceremonies differed across North America depending on what tribes were in each district, but there seems to have been a fairly universal association between rattlesnakes and rain. The Shawnees thought that thunder was the voice of a giant rattlesnake that lived in the sky, the Sioux felt that lightning was the same celestial rattlesnake striking its prey, while the Micmac said that black clouds were seven flying rattlesnakes which sped across the sky, calling out as thunder, and when they wished to return to earth they did so as lightning. One of the most attractive rattlesnake tales of all is found in the legends of the Kickapoo Indians of Michigan and Wisconsin who insisted that sacred rattlesnakes were responsible for rainfall, and told a story of a beautiful girl, the sister of an irascible rattlesnake who used to go around biting people. The girl was gentle and loving and whenever her brother bit someone she would follow behind and cure the injury. This only increased his frustration and anger, and in his fury he bit her and she died. But Manitoah-ah had been watching what had happened, and he turned her into a plant, the Arrow-leafed Violet, which the Kickapoo use in the treatment of snakebite.

The Huichol Indians of the southwestern part of the country believe that a rattlesnake god brings rain and is responsible for the growth of flowers as well as for taking all the children of the tribe into his care.

Quetzalcoatl, the Mexican god, had close associations with birds and rattlesnakes. Today the emblem of modern Mexico depicts an eagle with a rattlesnake (Lee Ironside)

But without doubt the best known of all the American Indian snake stories must be the Hopi Rain Dance, which is still performed, though it is almost certainly done these days for any tourists with wallets full of dollars in their pockets. According to the religious belief of the Hopi, the rattlesnake is a bringer of rain, so that at a time of drought a very necessary ceremony is performed that continues for nine days. On the first day the hunters set out to look for snakes. Before their departure the priest tells them, 'If you find a rattlesnake anywhere, pray to it and it may be raining soon. The crops will thrive and our children will thrive.' The hunters then depart. Though they will collect any species of snake that they find, they specialise in rattlesnakes.

It is not easy collecting snakes, it takes much time and patience; but at noon of the ninth day all the snakes are washed and taken to the shrine where dancing and singing ensues. As the ceremony continues the dancers become more and more wrapped up in what they are doing, and the snakes are draped about their bodies and even held in their mouths while assistants with feather whips distract the reptiles with gentle strokes. Finally a circle is drawn on the ground with cornmeal and the snakes are placed in a heap in the centre; they are then let go to carry prayers for rain back to the underworld when they return to their holes.

Researchers have shown considerable interest in the herpetological aspect of this event, and controversy has raged for many years as to whether or not the Indians use snakes that are able to bite, or whether they doctor them in any way. Shortly after the snakes have been released, some investigators have caught up numbers of the rattlesnakes to see what they could discover. In many cases it has been found that all the fangs, both those in use and all the reserves, have been removed with a knife, but some recaptured snakes had not been tampered with in any way.

This common worship of snakes extends down to Mexico where the Opata Indians have a similar belief about rattlesnakes and rain. Ophiolatry in the region goes back a long way. Quetzalcoatl, the Mexican god, is literally *quetzal* (a resplendent red and irridescent green bird from Central America), and snake, and much Central American art depicts snakes in religious and ceremonial situations. In the sixteenth century the Spaniard, Bernard Diaz del Castillo, paid a visit to Montezuma's zoo and said of it:

They have in that cursed house many vipers and poisonous snakes which carry on their tails things that sound like bells. These are the worst vipers of all and they keep them in jars and great pottery vessels, with many feathers. There they lay their eggs and rear their young, and they give them to eat the bodies of Indians who have been sacrificed.

When news reached Europe that in America there were snakes, variously described as bearing either bells or castanets on their tails, there was an immediate wave of horrified interest, and for many people rattlesnakes still seem to have an aura of menace that few other reptiles possess. When the first European settlers arrived in America, they too were to find themselves reluctantly attracted and repelled by rattlensnakes. The rebels in the American revolution carried flags depicting them, and the flag of the US navy consists of a rattlesnake stretched diagonally across thirteen red and white stripes, with the motto, 'Don't Tread On Me'.

Some of the new settlers took to snake worship, and the most famous of these new religions is the snake cult started in Grasshopper Valley, Tennessee, by a farmer named George Hemsley. His justification was to be found in the gospel of St Mark 16: 17-18 which states:

Believers will be given the power to perform miracles. They will drive out demons [sometimes translated as 'serpents'] in my name. They will speak in strange tongues and if they pick up snakes or drink any poison they will not be harmed. They will place their hands on sick people, who will get well.

At first the only members of the cult to handle live rattlesnakes were a few devoted church elders, but as the sect grew, more and more people joined in until even children were being involved in religious rites in this way. The cult spread to Kentucky in 1934, and when news reached the more rational outside world folk were horrified to find small children casually handling venomous snakes. At one time the members of the cult were content for outsiders to come and watch their ceremonies, but as pressure was put on them to refrain from allowing children to take part in these dangerous operations, they became far more secretive.

The silliest things have been done in the name of the Lord by followers of this church. In 1968 for instance, one of them, a minister of the Holiness Church of God in Jesus' Name, which is the cult's proper title, decided at a ceremony held at Big Stone Gap in Virginia that he would hold two large rattlesnakes against his temples. Both bit him and he died next day refusing, as do all the brethren, all aid apart from the power of prayer. In time it became customary for the faithful to have to undergo trials of faith by fire, and eventually some of them, taking the words of the Bible absolutely literally, took to drinking neat strychnine. Every one of them was soon with his Maker.

Unlike the more pragmatic Hopi Indians, these worshippers never took any precautions such as the removal of fangs before a session of snake handling, and researchers have been trying to discover why accidents were not more common. It was found that most bites occurred when the snakes were being removed from their boxes, which is not surprising. Once they were out, the constant movement up and down and round as the chanting and dancing progressed, tended to disorientate the snakes so that they were less willing to bite. In fact the movements were so extreme that very often the snakes went into a state of cataplexy. Nowadays such activities are illegal and, though the practice continues, the membership of the cult is getting smaller all the time.

It is not difficult to understand how venomous snakes became objects of veneration. They are secretive in their habits so they are not often seen, and they appear and disappear as though by magic, though in reality what happens is that they dive into holes at any disturbance or when it becomes too hot or too cold. They must certainly have seemed mysterious also in the way they caused death. After all the wound was minute, and probably often overlooked. When a man died in battle, or a hunted animal was killed for food, there was much blood sloshing about in all directions; but here was an animal able to cause death without any blood.

5·Venomous Myths and Legends

Legends and myths and misconceptions about venomous animals are legion. More rubbish is talked about these much maligned beasts than anything else in the world of nature, though stories about venomous plants are fairly startling at times.

A symbolic relationship between blood and snakes is the basis of many myths about them and is the reason why women were often regarded as having some sort of dubious connection with these reptiles. Generally a copious flow of blood resulted in the death of a victim, but at certain times of the month a woman could bleed as though badly wounded and yet apparently suffer no harm, and it happened again and again. So, since snakes caused death without blood, and women bled without dying, to the mind of primitive man there was an obvious link between the two, and this gave rise to many snake/women myths. In the Lebanon it is said in all seriousness that the shadow of a menstruating woman will cause a snake to become immobile. Of course it does; if a snake is wandering across a garden and a woman approaches so that the snake detects her footfalls, and then her shadow falls across it, it will freeze in the hope that it has not been noticed. But it will also do this if it is approached by a man, a donkey or a herd of gazelles. On the other side of the world, the Yuki Indians of California are convinced that the constituent of rattlesnake venom that actually causes death is menstrual blood, which they consider to be the most deadly poison of all. Since there is definite evidence that the blood of venomous snakes does contain venomous elements, this is an interesting theory. Further south in Argentina the Indians say that if a woman who is menstruating steps over a snake, it will die, and menstruation is so dangerous that any woman foolish enough to cut her hair at this time must take great pains to burn every last scrap of it for any fragments that fall to the ground will turn into venomous snakes.

The Hopi tribe, and the Navajo and the Zuni all say that the smell of a woman who is menstruating is so offensive to venomous snakes that they will make a point of hunting her out and killing her. In Europe the

Greek and Italian peasants even to this day warn their teenage daughters not to wander through the vineyards when they are menstruating as this is a sure way of attracting venomous snakes. In Australia it is believed by the Aborigines that if a woman goes at this time of her cycle to a waterhole, the snake that guards it will abandon it and move to another area, which will result in the waterhole drying up completely, a catastrophe that must have been blamed on many women over the centuries.

One of the strangest tales of all is a legend told by the Kwakiutl Indians of the Pacific coast of America about a supernatural woman whose vagina is surrounded by sharp teeth, which is synonymous with the mouth of a rattlesnake.

If you want to enjoy a really good laugh and learn any number of myths about snakes at the same time, do try to read some of the books by Victor G. Norwood which were written in the 1950s. They have long been out of print but are worth searching for in second-hand bookshops. Victor Norwood was an adventurer who spent much of his life in South America, earning a precarious living by panning for gold and exploring, and picking up what money he could here and there along the way. Inevitably much of his life was spent miles from anywhere, deep in the rainforest of Amazonia or, as he would much prefer to call it, 'the green, stinking hell of the South American jungles'; he went in for that sort of purple prose. His life was interesting and clearly dangerous, but he felt he had to embroider it somewhat to please his readers, or perhaps his publisher!

Snakes come in for more embroidery than anything else and he tells marvellous stories about them. On one occasion he and some companions were travelling along a small river somewhere in Amazonia in a canoe, and just as they approached a bend they became aware of a horrendous din round the corner. When they could see the cause of the disturbance they found a calf standing on a small, flat, grassy area beside the water. From the river itself emerged a huge Anaconda that had sunk its teeth into the neck of the bawling animal that was pulling backwards with all its strength, for the snake had clearly anchored its tail to something beneath the surface of the river. The calf was pulling so hard that the reptile was stretched as tightly as an elastic band, and quivering with the strain. Upon seeing the plight of the poor old calf, Norwood stepped forward and with a mighty slash from his machete, sliced straight through the Anaconda about 4m (12ft) from the head. Like a snapped hawser, both halves whipped away, the back end to disappear beneath the water while the front let go of the calf and was flung across the grass. The snake glared intently at the man and started to approach menacingly, keeping its eyes on him the while so that he was unable to move. Soon it was close enough for him to feel its breath on his face, and had it not been for one of his colleagues who shot it through the head and killed it, Norwood would surely have died.

The idea of the Anaconda stretching like an elastic band is a popular misconception. In reality a snake is an animal much like the calf in the story. It has a heart and lungs and muscles in just the same way, so if you were to cut one in half, the two parts would drop to the ground and writhe. The idea that snakes are able to hypnotise their prey by keeping their eyes firmly fixed on it is also a myth, though it is one that has been repeated throughout history. What actually happens is that a snake, when it is approaching a potential meal, does so very slowly so as not to draw attention to itself. At the same time, it keeps its head firmly facing the direction of the victim so as not to lose track of its whereabouts, and a snake cannot blink since it has no eyelids. While all this is happening the animal that is about to be eaten is often aware that something is going on but is not sure what, so it sits up and peers hard in the direction of the threat. Each time it relaxes slightly the snake makes an infinitesimal movement, and that again attracts the animal's attention, though since there is no sudden movement, there does not appear to be enough cause for alarm and flight. The two animals can spend quite a lot of time in these relative positions, which must have given rise to the story about hypnosis.

Just as persistent is the tale of the snake's poisonous, or at any rate foul smelling, breath. A snake's breath does not smell of anything much, though if one were to try and sniff it while the snake was just about to regurgitate food as a result of fright, it would probably smell unpleasant. It certainly is not venomous though it is easy to see how the suggestion arose. A snake will often strike so fast that it is virtually impossible to see what happens. After the strike it reverts to a typical defensive position, with its head well back, ready to strike again, and a good S-bend in the neck to give it reach when it needs it. Thus it immediately resumes the position it was in before the strike, so when a victim suddenly starts to exhibit symptoms of snakebite without anything apparently happening, what other explanation could there have been in the minds of earlier generations than that the snake had killed someone merely by breathing on them? Another possible reason for this story seems to have originated during the Punic Wars (264-146BC), when soldiers reported a strange creature that they called a Basilisk, which could kill a man with its venomous breath. Nowadays it is thought that the Basilisk was probably a Spitting Cobra, which the troops came into contact with

Popular books often show venomous motifs on the cover (John Nichol)

for the first time. This venomous Basilisk should not be confused with a South American lizard of the same name that has the delightful habit of racing across the surface of water so fast that it does not sink. It is not in any way venomous.

One of the commonest of all snake stories is that no snake will die until sunset. If you chop a snake's head off at ten o'clock in the morning, it will die at ten o'clock in the morning, but snakes are able to put up with an astonishing amount of physical abuse before they die. It could easily happen that a snake was attacked during the day and left by someone who was convinced it was dead, when it was only mortally wounded. It could still have been alive when its attacker went to have a look at it again in the evening after he had finished his day's work. Next morning when he looked again it had perhaps finally died, and thus was such a myth born.

One of the funniest of all snake stories, told of many species but especially of cobras, is that they drink milk. Snakes are remarkably reluctant to drink milk though they might have a go if they are very thirsty and there is no water available, but still the story persists. You even hear tales of people who insist that they have fed a snake on milk for a long time. In that case it is either finding food elsewhere, or it is a well-fed specimen kept in a small enclosure where it cannot move much. A snake can live for many months, perhaps even over a year in the case of a very large one, without feeding. Their metabolic rate is slow and a meal will last them for ages. Consequently a snake that is not fed at all, or one that is given only milk, can live for a long time before it succumbs. But the story about reptiles drinking milk is so firmly believed that the tale has been extended. It is said that snakes will climb the leg of a cow to drink directly from the udder and will even hunt out a lactating human female and wait until she is asleep so that it can suck from her breast. All this would be totally absurd except that evidence shows rattlesnakes are attracted by the smell of rotten milk and will come from some distance to investigate the smell. Funnily enough they do not drink it when they arrive so nobody really knows why this happens.

Throughout history people have been saying the silliest things about snakes. As far back as the 4th century BC, the legionnaires of the Macedonian army were coming home and enthralling their families with stories of snakes with eyes as big as shields. Cleopatra is alleged to have killed herself by getting an Asp to bite her, but this seems an unlikely death to choose since an Asp is a viper, and as we have seen viper bites are horribly unpleasant. It is far more likely that the snake she used was an Egyptian Cobra, for although not many people would choose to die by snakebite, one that injects a neurotoxin is less horrible than one that employs a haemotoxin. Explorers and soldiers who had spent time in exotic parts of the world were constantly bringing back strange tales of the animals they met, and even during the Vietnam war US marines talked of a snake whose bite was so virulent that it would kill before they could walk six paces after being bitten. (Only recently somebody who had spent her early years in the Middle East told a story about 'camel spiders' in the desert which crept up on sleepers and injected some sort of liquid that anaesthetised the bite so that the animal could then eat a chunk of tissue from the site of the wound without waking up the victim.)

Throughout history venomous snakes have been used as an efficient way of killing people and there are many documented instances of individuals being offered an honourable death by snakebite rather than beheading. Death by snakebite was a favourable form of execution in parts of southern Europe at the end of the sixteenth century, particularly for the crime of parricide, and bites by a variety of venomous snakes have been used as a form of torture to extract information from the unfortunate victims. Snakes have also proved useful as instruments of war. As long ago as the 3rd century BC the Carthaginians defeated the Romans at sea by catapulting earthenware pots of living snakes onto the decks of their ships. Without any effective means of treating snakebite this must have been awfully effective, and would undoubtedly have thrown the Roman crews into panic. Similar practices have continued, and even during the Vietnam war snakes were being used as booby traps by the Vietcong. There is such horror and revulsion among people about venomous snakes that the psychological effect must have far outweighed the damage they did.

Murderers, both in real life and in fiction, have found venomous snakes useful ways of getting rid of their victims. The likes of Sherlock Holmes and James Bond are forever having to fight them off and, in the middle sixties, civil rights workers in Mississippi who were supporting rights for negroes sometimes found that rattlesnakes had been placed in their cars. Reprehensible though this is, it has a sort of crazy logic; after all, if you want to kill someone, to a layman this must seem an effective way of achieving your end. What seems totally incomprehensible is the habit of putting rattlensnakes in the beds of friends as a practical joke. One of the saddest stories of all must be that of a Californian man who wanted to murder his wife. Much planning and thought went into his scheming and finally he decided that he would try and kill her by using a rattlesnake; that way the woman's death could appear to be an accident and he would get away with the crime. So one day he placed a snake where she would be likely to be bitten and, sure enough, the reptile sank its fangs into her. But as we have already seen, nothing is definite with snakebite, and though she received a good dose of venom and suffered horribly, she recovered from the bite, so he drowned her.

Rattlesnakes have always been credited with supernatural powers of one sort or another. In the days when the pioneers were trekking west across America, it was said that one had to be careful at night when the covered wagons were moved into a circle and the horses unhitched, for if you let the shaft drop onto a rattlesnake the animal would bite it with such effect that over the next few days the wood from which it was made would become rotten until it finally crumbled away. The rattlesnake was always blamed, and perhaps no one thought that the wood might be rotten anyway. Another story that originates from about that time, and for a change is actually true, is that if you chop off the head of a rattlesnake it can still bite you. This does not apply only to rattlesnakes. The head of any snake may close its jaws after the animal is dead, and if the fangs enter a finger an injection of venom may very well be the result. It has been found that this could happen as much as twenty-four hours after death, but even long after that it would be unwise to play with the fang of a dead venomous snake in case you prick yourself.

One of the most common questions is what is the most venomous snake of all. There is no easy answer since the question can be interpreted in many ways. Does the questioner mean, for example, which snake kills quickest, or which species is responsible for most deaths, or which individual snake has the most venom or perhaps the most toxic venom, or even which species is most likely to bite. In Pakistan the inhabitants have no doubt that scientific claptrap about cobras and Russell's Vipers and so on is just so much nonsense. They insist that by far the most dangerous snake in the country is the dreaded Hunkhun. Not only does it perform all the horrific functions that a venomous snake is supposed to, when it is struck with a stick its body breaks up into many pieces, and when it is wounded each drop of blood is just as deadly as the venom. Some years ago there was a virulent correspondence in a Pakistani newspaper, each writer trying to outdo the last with tales of the Hunkhun. When asked to identify it everyone points to a gecko, a small, perfectly harmless lizard. The only venomous lizards in the world are the Gila Monster and the Beaded Lizard from the southern states of the US and northern Mexico, but despite this there are stories of deadly lizards throughout the world. Admittedly some of the bigger species can give an excruciating bite if you mess about with them, but they are not venomous. The Jogis of Pakistan, mentioned in the last chapter as a tribe of

Any children's party will find someone dressed as Miss Muffet, who was a real person. Her father fed her spiders to cure her childhood ills (John Nichol)

snake handlers, are competent herpetologists. They are also businessmen, and they will happily sell a potion to make gullible customers immune to the effects of the venom!

In Senegal, in west Africa, it is said that Puff Adders have a ball of grease at the base of the tail impregnated with all sorts of magical powers. These powers are very useful to the snakes, but for some reason it does make them sensitive to the smell of human faeces which they hate so much that they will jump over ten houses to avoid it. If some-one should be so foolish as to touch one with a stick that is smeared with this substance, the snake will be so offended it will chase him for hours, even if he should be on horseback, until finally it catches him and kills him. There are all sorts of stories about the speed at which snakes can move, and they have been accredited with phenomenal rates. The Black Mamba is said to be able to outpace a man on a motorcycle, and is generally accepted to be the fastest of all snakes, which it probably is, but not nearly as fast as that!

The hunters of the Efik region of Nigeria try not to kill venomous snakes if they can help it, but if this should accidentally happen they put the body of the animal across the path until the next new moon. If they neglect to do this they will be bitten by another snake. It is not clear whether this is in revenge or just coincidental, but the revenge theory is invariably attributed to the King Cobra. Kill one, and the mate will always hunt you out and kill you in return. In the Rudyard Kipling story, 'Rikki Tikki Tavi', the same is said of the common Indian Cobra.

Anyone who handles reptiles is immediately credited in some parts of the world with the ability to charm snakes. A traditional charm against Adders is common throughout Britain and northern Europe. One has to draw a circle around the snake and recite 'Let God arise, let his enemies be scattered, and let them also that hate him flee before him. As smoke is driven away, so drive them away; as wax melteth before the fire, so let the wicked perish in the presence of God.' In real life, Adders are so unlikely to do you any harm that one wonders why bother with all this rigmarole.

Almost all the myths and legends relating to venomous animals refer to snakes, though a few concern spiders. Probably the best known of all is the tale of Miss Muffet. The nursery rhyme tell us that

> *Little Miss Muffet sat on a tuffet*
> *Eating her curds and whey*
> *Down came a spider*
> *And sat down beside her*
> *And frightened Miss Muffet away.*

Miss Muffet was a real person. Her name was Lucille Mouffet and her father, Thomas Mouffet, was a doctor who lived in France during the last century. At that time there were numerous country remedies for all complaints. Most of them incorporated plant material of one sort or another, and some involved animals; slugs, eaten live and whole, were reputed to cure lung problems. Doctor Mouffet was a great believer in the efficacy of spiders for a range of common complaints. Therefore whenever poor

Lucy contracted any one of the host of illnesses to which small children are prey, her father would dose her with live spiders. According to the story these were sometimes smeared on toast with a knife. It is hardly any wonder, therefore, that the little girl developed the intense dislike of spiders which accounts for her running away when one came and sat down beside her. Though there is no real evidence that a daily dose of spider is of any real use, spiders' webs have been used effectively for many hundreds of years in the treatment of wounds. A piece of sheetweb laid across an open wound is very good in arresting bleeding; blood clots faster when the mesh of the web is in place over the injury. The reason for this is purely mechanical, but it works and was still being used in Britain a generation ago in country districts.

Spiders are universally regarded with disgust, at any rate in the West, but at the same time, the so-called 'Money spiders' are regarded as lucky. They are too small to be perceived as any sort of threat and their presence is said to mean that one is about to receive a sum of money from an unexpected source.

The best known scorpion story claims that if one surrounds a scorpion with a ring of petrol or other inflammable substance, and sets light to it, the scorpion on finding itself unable to escape will sting itself to death. It is an indication of man's detestation of venomous animals that this theory should ever have been put into practice, but it was a common diversion of Allied troops in the deserts of north Africa and in India during World War II. As with many of these myths, poor observation of what is actually happening is the origin of the story. A scorpion in such a situation does not really commit suicide, but just as any other animal being burnt to death twists and contorts as tissues are seared and fluids expand, the abdomen and tail of the scorpion curls tightly over the back until it looks as though the sting on the end is being used to inflict death.

Many of the myths to do with venomous animals are directly related to means of acquiring immunity or of alleviating the effect of a bite or a sting, and in Greece the peasants until fairly recent times used to catch a live scorpion and drop it into a small bottle of olive oil. The bottle would be corked and the animal would eject a thin trail of venom into the golden liquid before it died. The bottled scorpion was kept as a safeguard against the sting of one of these animals. In the event of such a thing happening, the idea was to rub some of the oil from the bottle onto the affected part to remove the pain and swelling. The only other similar case is to be found in China where a live gecko is dropped into a bottle of rice wine before bottling. Geckos are not venomous, and it is not clear what this is supposed to achieve unless it is to improve the flavour, but all the bottles of certain sorts of wines come complete with dead gecko.

Toads are venomous, perhaps more venomous than many realise, so it is hardly surprising that they were long associated with witches. Together with black cats, toads must be their most common familiar. This reputation probably started when witches incorporated toads into their various mixtures in order that the venom should form a constituent of potions designed to remove unwanted enemies. The three witches in Macbeth give comprehensive details on how to prepare such a brew. The ingredients are amazingly complicated and include toads:

> Round and round the cauldron go;
> In the poison'd entrails throw.
> Toad, that under coldest stone,
> Days and nights hast thirty one
> Swelter'd venom, sleeping got,
> Boil thou first i' the charmed pot.

Shakespeare was clearly familiar with the venomous aspect of toads and the process of extraction. One would expect him to be, for on reading through his work it is obvious that he was a pretty good naturalist. Presumably the reference to the toad sleeping beneath the stone for thirty-one days refers to winter hibernation.

Witches must have collected the toads for their brews in the spring when the animals first emerged after the cold weather and made their way in thousands to their spawning ponds. The end of April coincides with one of the major festivals in the calendar of ancient religion, or devil worship as some would call it. Consequently it must have appeared obvious to simple folk that a witch was collecting the amphibians to take to the ceremonies. It is certain that the witch must have kept some of the toads for use later in the year, so the story of them as witches' familiars is easy to follow. Given that they have a creepy image it seems contradictory that there is also a tale which reveals that if a young maiden kisses a toad, he may turn into a handsome prince and carry her off on his white horse to a castle with a multitude of spiky turrets. Witches apparently had a habit of turning such princes into toads.

If you read through the complete spell that the witches in Macbeth so kindly provide the recipe for, you will discover that the ingredients also include 'Adder's fork, and blind worm's sting'. It is strange that the fork, which presumably means the forked tongue of the Adder, should

Bee larvae are fed on royal jelly. It is also sold in shops as a health-giving product (John Nichol)

be used. The tongue is harmless so one would have thought that it would have been more sensible to use the whole head, thereby incorporating the venom glands as well. The blind worm sounds as though it could be a Slow Worm, but the latter has no sting. A Slow Worm is a lizard and quite harmless. Apart from that it is not blind; it has perfectly good eyes. The end of the tail is hard, so it could be thought to be a sting, though it is not sharp. If on the other hand the animal referred to is an earthworm, which is blind, surely nobody could ever have attributed any sort of sting to it. They must have had good witches' supplies shops in those days otherwise where would they have obtained 'Tiger's chawdron' and 'baboon's blood' in Macbeth's time?

Salamanders are rather like the fabled Phoenix. Throughout their range in Europe they have had a reputation for being born from fire. The reality is that these delightful, handsome black and yellow animals tend to spend their time hidden away in any suitable dark spot except in spring when they are breeding. Obvious places to hide in were hollow branches lying on the ground and when these were put on a fire by our ancestors and the salamander began to feel uncomfortably warm, it would wriggle out of the branch and walk away from the fire. For this reason it acquired its common name of Fire Salamander.

Myths and legends extend to insects too. Bee larvae are fed on a whitish liquid made from a mixture of pollen, honey and various secretions from the glands of nurse bees. This mixture, which is very rich in protein, is known as royal jelly and a small industry has evolved supplying royal jelly to wealthy, ageing ladies as a rejuvenating compound. But despite the many advertising claims for the product there is no conclusive evidence that it does any more good than other products claiming to make them more youthful. Some years ago there was interest in a treatment for arthritis and related conditions using the venom of honey bees which were induced to sting a patient around the infected joint. Delighted patients, (or were they victims?), subsequently claimed their aching limbs felt infinitely better after the treatment; but however bad your arthritis, if you suffered from half a dozen bee stings into the joint, would you not feel better when the effect from them had worn off?

6 · Snake Charmers

Snake charmers have always fascinated man for the apparent power they have over strange and potentially lethal animals. In essence the act of snake charming is simplicity itself. The charmer tells his enthralled audience that he is going to make the snake dance to the music of the flute. As we saw earlier, a snake cannot actually hear. It is true that it can pick up vibrations through the ground, but that ability plays no part in a performance. Snake charmers usually have about half a dozen snakes which they use in the course of their act. Most of these are cobras since they look so impressive with their hoods spread, but there is usually a python or two in a basket, and occasionally a couple of other non-venomous snakes. It is not often that one sees vipers used by snake charmers. The snakes live all day coiled in a tiny basket. When the lid is opened the snake leaps upright startled by the bright light, and completely disorientated. While it is still trying to adjust to the new situation the charmer starts to play his flute, moving it constantly within a few inches of the snake's face. The snake sees the end of the flute as a potential threat and moves to follow it so that if need be it can strike. It is this movement that the snake charmer describes as the dance of the cobra.

These snakes, the stock in trade of the snake charmer, lead a pretty miserable life. Most performers take precautions against being bitten and the first thing done with a newly caught cobra is to remove the fangs. These are usually cut out with a knife, or in some parts of the world the snake is induced to strike again and again at a stick which has been wrapped up with rags. As the snake strikes, the stick is snatched away, pulling the teeth from the snake's jaw. Some people say that the lips of the snakes are stitched together, but if that is still the case, it is only done rarely. Whichever method is used to render the snake harmless, it is further trained by bullying it mercilessly for several days. It is hauled around and slapped and prodded so that it gets used to being handled and to make it reluctant to strike, which in any case it is not keen on doing since its mouth is sore and sensitive. By the time the reptile is considered ready for its debut, though it can strike it is very unlikely to do so, and in

Cobras feature predominantly in the design of the board game Piremid, based on life in Ancient Egypt (John Nichol)

fact if you watch a performance you will see that although a cobra will sometimes duck its head in the direction of a hand or the flute, it will rarely open its mouth to strike. It is only a half-hearted bluff.

These snakes are in poor condition. They are often suffering from heavy loads of external parasites and are frequently seen to have scars and damaged spines. Unable to bite they cannot feed normally either, and since they live in their baskets they have little opportunity to drink so they are likely to be dehydrated as well. It is true that each evening when he is able the snake charmer will take the snakes from their baskets and dunk them briefly in water, but it does little for their general health except perhaps to prolong their miserable lives a little. Within a few weeks or months they die. There are tales that some snake charmers treat their charges with reverence and after using them for some time release them again in the wild, but if there were ever any truth in this statement it must have been an isolated occurrence. Snake charmers just do not treat their animals like that.

The scope for entertaining an audience by playing a flute to a reluctant cobra is small, so part of the performance is taken up by hauling the animal around and trying to persuade him to erect his hood. After that he is stuffed back in his basket and it is the turn of a python or two. They don't do much either, but they look good draped across the shoulders of the charmer, and if he can persuade a tourist to hold the snake it not only adds variety to a performance but is also good for bigger tips from the audience. When this has paled – and the snake charmer soon spots when the watchers are getting bored, for his knowledge of snake psychology is only surpassed by his knowledge of human psychology – the python is put away and any other members of the cast are brought on view. Sometimes there may be a sand boa which has a blunt rounded tail, and the audience is told that the snake has two heads, one at either end so that it can travel in either direction. Sometimes a brightly coloured snake such as the magnificent black and yellow Mangrove Snake is brought out, or the Red-tailed Green Racer,

(top right) *The extraordinary camouflage of some venomous fish results in swimmers stepping on them. This handsome specimen is a Scorpion Fish*
(below right) *Despite their venomous spines these Weever Fish are on sale for food*

This yoga position is known as the Cobra (John Nichol)

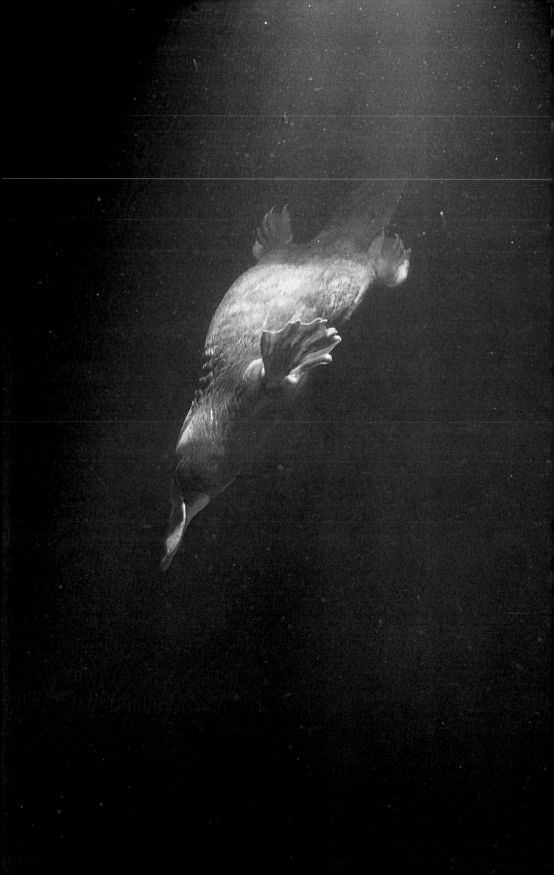

a lovely thing of painfully bright green and, as its name suggests, a red tail. These do nothing except look dramatic, and after they have been packed away all that is left is to pick up all the coins that the appreciative audience has thrown down, and leave.

The snake charmer does not have an easy life. He wanders throughout the country as a rule, sleeping wherever he can, earning little money and doing a potentially dangerous job. Although he takes pains to remove the fangs from big cobras he and his fellows do get bitten by venomous snakes, many of them frequently, and although this results in an immunity of sorts new bites are always painful. Charmers generally stick to cobras, which are usually placid. Some vipers are irascible, and even after the fangs have been removed there are always the reserve teeth ready to be brought into action. Even when cobras have been treated there is always a risk and you will see a charmer dodge strikes, not just for show, but just in case. Burma seems to be the only place in the world where there are women snake charmers, and they are renowned for the fact that their snakes are King Cobras, and that they end their act by kissing the cobra on the nose.

It must be a rare snake charmer that makes his living entirely by giving performances. Each of them earns a little here and there by selling snakes to collectors, dealers and snake farms for the production of venom for medical use. They also provide a service by clearing areas infested with snakes. They genuinely do this though there is an element of show business about it as well, and they usually have a few cobras secreted about their person before they start so that these can apparently be hauled squirming from suitable holes while the patron is watching.

Snake charmers always have their eye on ways of making money so most of them are also willing to provide a selection of charms, potions or amulets against snakebite. Famous amongst these are bezoars. For centuries there was much mystery surrounding these objects, which look like small, often highly polished stones. No doubt such mystery was furthered by the snake men themselves. Bezoars were supposed to be the best of all defences against snakebite, and the story was that they were found beneath the skin of a King Cobra's forehead. The truth is far more mundane. They are usually concretions from the stomachs of goats, or they may be made from the antlers of deer, or from bone. Less frequently they are kidney stones or gall-stones. The best ones are very,

(left) *The Duck-billed Platypus from Australia has venomous spines on its front limbs*

The world famous snake temple in Penang, Malaysia (Herbert Lieske)

very expensive, so much so that they are far too costly to be owned by an individual and are purchased by a group of shareholders. The theory behind the use of bezoars is that if you are bitten by a venomous snake you apply the bezoar over the site of the wound to 'draw out the poison'. Since we have already seen that the nature of snakebite is such that most victims will not die anyway, these placebos must seem good value. After use they are regenerated by boiling them in milk or some exotic liquid. There is no truth in the story and many of the things are not even absorptive - some of them, for example, are semi-precious stones.

Variations can be purchased which are said to protect the owner agains the stings of scorpions or even rabies. In the US they are known as madstones from the rabies connection, since that awful condition, hydrophobia, is supposed to turn the victim mad. In some parts of the world bezoars are known as muns. Amulets which claim to have the same effect are frequently called gris-gris and in Senegal, one of the many African countries where they can be purchased, they are called lars and are sausage-shaped pieces of wood bound with pieces of coloured cloth into which are inserted tufts of horsehair. They are bound to the calf of the leg.

In some parts of the world opportunists of all sorts take up snake charming as a way of earning a little money during the tourist season, especially today when long-haul holidays are becoming more commonplace for affluent visitors who would not know a genuine snake charmer from a cowboy. But traditionally charmers have come from families of snake charmers who can trace their roots back for centuries. The Psylli are the earliest known snake handlers. Their fame was great and for centuries they monopolised the trade in snake venoms and cures for snakebite. When Queen Cleopatra was bitten by her Asp, Augustus Ceasar called in the Psylli to try and prevent her death. Each tribe and family of Psylli continued in this profession, passing trade secrets on from one generation to the next. Eventually the world began to change; but the Psylli like all snake charmers were survivors and changed their religion to that of Islam. They exist today. A report dating from 1900 talks about their direct descendants, the Saadeyeh dervishes, commemorating the birth of the prophet Abraham by swallowing live snakes to give themselves immunity. The snakes' mouths had been secured with silver rings, and on the belt and braces principle the lips had also been sewn together with silk. Their spell for clearing houses of venomous snakes is still in use today, and was recorded virtually in the same form hundreds of years ago:

I adjure you, by God, if you be above or if you be below.
That ye come forth.
But if you be not obedient, die, die, die.

Either way they can't lose. If the snakes are obedient they come out to be caught, and if they are not they die in their holes.

Most of these charms to prevent or cure snakebite are pure folklore, but west African charmers have one that really does work. They make a pulp from the sap of two plants, *Euphorbia hirta* and *Ageratum conizoides*. They rub the pulp over their bodies and then stroke the snakes with it. Finally they insert a lump of the pulp into the snake's mouth. This induces tetany of the jaw muscles, and causes inflammation of the venom glands which renders them temporarily useless so that the reptile can be used in the performance with impunity.

Other communities of snake charmers also go back a long way and Pliny records that the Marsi, who lived near Rome, were descendants of the goddess Circe and that they were immune to snakebite. The present-day snake festival at Cocullo in central Italy can be traced directly, from the original Marsi. Pliny incidentally was unaware of the presence of venom glands and suggested that venom might be produced in the gall bladder.

A modern version of the shows put on by snake charmers is the reptile show staged in some states of America. Like their Old World cousins the snakes do not actually do very much, and considerable expertise is required on the part of the handler to give the illusion that anything is happening. It is essential to build up interest so that the paying customers are prepared to enter the premises to see the show, and the barkers outside talk about 'Killers from the jungles of the Amazon and the great peninsula of India, which feed on the warm blood of the living. . .' All their performers have had their fangs and venom glands cut out with a knife long before they ever appear before the public, and much of the snake charmer's work consists of trying to stir the snakes up so that they will actually do something. In the past such shows were often extended to provide greater thrills for the audiences. There were, for example, fights between rattlesnakes and king snakes, or even fights between rattlesnakes and dogs and cats; and for those with really jaded tastes there was always the chance to watch a 'geek' at work. Geeks were men who ate live snakes during the public performances. The fights between snakes and other species of animal bring to mind the cobra/mongoose battles that are surreptitiously offered to tourists in India even today, and just as the snake charmers made extra income for themselves by selling odds and ends, so too did these rattlesnake shows in the States sell such things as rattlesnake oil which was guaranteed to cure rheumatism.

None of these entertainments made any pretence at being educational or scientific, but some of the better zoos that believe that part of their role is to educate the public hold reptile-handling demonstrations periodically, and those that demonstrate the control of venomous snakes and the way they are milked of their venom are always sure of a good audience.

7·Venomous Animals in the Modern World

Venomous animals enthral us so much that our culture has become filled with symbols and references to them which we take for granted. There is not a single day when some mention is not to be found in the press. For example many newspapers carry astrological predictions and one of them, Scorpio, is the eighth sign of the zodiac and covers the period from 24 October to 22 November. Though many people profess not to believe in it there can be hardly anyone who has not at some time looked up his stars for the day to see what is in store. Furthermore the

Scorpio, the star sign, whose attributes are said to represent all that most people find detestable in venomous animals (Michael Beasley)

interest seems to be increasing. One astrologer is quoted as saying that the twentieth century will be remembered by future historians as being the time of the great renaissance of the science. Professional astrologers state that for a forecast to have any accuracy they must have the exact time of a subject's birth, so of necessity the forecasts in the daily newspapers cannot give too many details for readers about how their day is going to go. In fact they are impressive in their vagueness!

Scorpio people are said to be, amongst other things, secretive, wilful, obsessive, destructive, jealous, possessive and caustic; but lest the Scorpios reading this book tear it up in rage, it must be said that the textbooks on astrology also list numerous attractive attributes of those born under this star sign. It is interesting, however, to see the sort of words used, when one remembers the feeling most people have about venomous animals. Astrologers say that Scorpios can be secretive, and in sinister fashion seek to manipulate others. This particular sign is seen as evil by those who are uneasy about the unknown, and in-depth analysis of Scorpios often uses spiders's webs, serpents, eagles and actual scorpions as symbols to make points about intelligence and other characteristics. According to the literature, Scorpios, when looking for careers, should examine fields that are particularly appropriate for them, including sexology, mortuary science, sewage and disposal work, and slaughtering!

The history of astrology as we know it in the Western world demonstrates that all the cultures whose mythology contains tales of scorpions, cobras, toads and other venomous animals, have had a hand in the history of the science. The Ancient Greeks, the Aztecs, the world of Islam, and all the rest have a similar interest in astrology. The Chinese developed their own version. Instead of the symbols that we are familiar with they use twelve animals, each of which rules the fortunes of a given year in turn; 1989 is the year of the snake. The ox, or bull, is common to both sets of symbols, each of which contains a single venomous representative; for though only a small percentage of snakes are venomous, the Chinese snake symbol is usually shown as a viper of some sort. Horoscopes from the Chinese almanacs make delightful reading. They tell us that people born in the year of the snake are endowed with tremendous wisdom, they are profound thinkers and rely on their own intellect rather than trust the judgement of others; fired with intense determination they hate to fail at anything. Snake persons seldom have to worry about money as they are usually rich, and yet they are stingy and will not make loans, even to friends. Snake men are handsome and the women are beautiful, and both are inordinately vain. They are said to enter into extramarital relationships frequently and with enthusiasm,

A wide range of products uses names of venomous snakes (John Nichol)

Toys representing venomous animals are common, and one can even buy edible tarantulas (John Nichol)

with the result that they tend not to have happy marriages. Finally, we are told, snake men and women have an annoying habit of overdoing everything, including helping others!

Have a look in your local telephone directory to see how many venomous entries you can find; you will be surprised how often such names crop up. The London directory lists twenty-one companies with the name Cobra, and there are various King Cobras and other serpentine titles; these are only exceeded by the number of Scorpios or Scorpions. There is also a scattering of Tarantulas, Spider's Webs and a couple of Adders. Even where a company is not given a venomous name, many use such titles for some of their products, and these are particularly common in the world of aviation. Amongst the fixed-wing aircraft there are the de Havilland Venom, Bell's Airacobra and Kingcobra, McDonnell-Douglas's Hornet, and Northrop's Scorpion and Black Widow. There are also Cobra, Hornet, Scorpion and Wasp helicopters, and some aircraft are propelled by Pratt & Witney's Wasp and Hornet engines, and Rolls Royce's Vipers.

The whole business of killing people is full of such imagery, for in addition to the military aircraft, there are Sidewinder missiles and Stingray torpedoes, and the United States navy has aircraft carriers named USS Hornet and USS Wasp. One can understand why such names are chosen for powerful, thrusting lethal bits of machinery, and this sort of thinking has led to toys and playthings of a warlike nature. Any toyshop will sell you a Scorponok should you want one – a space vehicle that resembles a Scorpion, and the range of Mask toys, who are apparently spacemen of indeterminate sort, battle constantly with the evil forces of Venom. Not everyone over the age of eleven has heard of Scorponok or Venom, but they must surely be aware of Spiderman. This character of Marvel comics has become so popular that cartoons about him fill our screens, small boys dress up in Spiderman suits, and Spiderman merchandise proliferates year by year. New Spiderman comics are constantly coming on the market, and the whole operation of writing and merchandising the character is a highly sophisticated business.

There is even an official biography of the character which makes fascinating reading. According to this document, Peter Parker, a freelance photographer and 'adventurer' from New York, attended a demonstration on the handling of nuclear waste. An unfortunate spider happened to wander into the scene and suffered a massive dose of radiation but, before expiring, the spider bit Peter Parker on the hand. The venom from this irradiated spider turned him into a sort of super person with tremendous strength, the ability to climb vertical surfaces and a host of other useful attributes. Not surprisingly our hero decided to earn himself a living through his new powers, making himself a suitable costume, and calling himself Spiderman, he appeared on television and was soon a media personality of even greater stature than Terry Wogan. Later still his uncle was killed by a burglar, and, full of remorse because he felt he could have prevented the murder, Spiderman turned instead to a life devoted to combating evil. This he has achieved successfully for a number of years, during which time he has accumulated a number of useful gadgets to help him. He now has an arsenal of web shooters, special cameras and various other bits and pieces to help him in his endless fight against crime. Spiderman has caught on with succeeding generations of small boys to the extent that one can almost guarantee that any children's party will have its contingent of mini Spidermen.

Venomous animals in the form of toys are clearly popular with children, and the variety of creepy-crawly toys available is truly astonishing. There cannot be a toyshop in the country that does not have a box full of horrible hairy spiders made from clammy rubber, and even some adult men buy these things to frighten the women in their lives. The proprietor of a shop that specialises in tricks and jokes commented that he had never known a woman to buy any unless accompanied by a child in the shop who asked for them. Most of these joke animals are spiders or snakes, but by careful hunting one can find scorpions, octopuses, ants, and even centipedes. One supplier of such playthings says that he imports them by the container load from Hong Kong, and if one cares to browse through the catalogues of the various products that are produced in that enterprising place, one can see that a number of firms make them. Strangely, these animals are all designed to cause a mild shiver of disgust or fear, yet bee toys and ornaments are usually never regarded in the same light. Instead they are fun things, furry and brightly coloured. They are almost invariably depicted with huge eyes, and sometimes a wide smile to appeal to us as cuddly little friendly beasts. In real life many people go mad whenever a bee appears in their vicinity.

Human beings are not born with an innate dislike of venomous animals, but the general revulsion is learnt at an early age. Primary school teachers who should know better tell their small charges before they are out of their first year that snakes are horrible slimy things, and with this sort of example it is hardly surprising that children grow up to hate and fear these animals. The venomous toys clearly provide the same sort of horrible thrill as does a ride on a Big Dipper or other fairground ride. What does seem surprising is that when a young boy discovers that his little girl friend is frightened by spiders and buys one to frighten her, she actually goes along with the charade and runs off squealing at what she knows is a harmless toy.

When one considers this upbringing and the fact that numerous expressions relating to venomous animals have found their way into our language, it is hardly surprising to find our books and films packed to bursting with spiders, scorpions and the rest. In English we talk of a 'viper in the bosom', a 'waspish tongue', a 'web of deceit' and a 'snake in the grass' – the language is full of such terms, all of them referring to people or patterns of behaviour generally regarded as antisocial or reprehensible. Given therefore our obsession with these animals it should not be a surprise to find many fictional books about them. Some of these stories are written as straight thrillers; others are pure hokum, and not meant to be otherwise. *Venom*, a tale about a small boy who buys a pet snake only to find that he has been sold a venomous one by mistake; *The Speckled Band*, a Sherlock Holmes adventure by Conan Doyle, or *Octopussy*, a James Bond story by Ian Fleming: all make use of the venomous aspect of the respective animals as part of the story.

On the other hand there are books and films named after venomous animals which have nothing whatsoever to do with them, such as Cobra, a tale of a tough policeman from Los Angeles who hunts a serial killer. Black Widow was a popular title for film makers and there have been four completely different films of the same name. The first, made in 1947, concerns the daughter of an Asian king who is sent to the US to steal the engine of a rocket. The second film (1951) tells of a man who kills the enemy who tries to murder him only to find that his wife wants to take advantage of the situation by claiming that the body is that of her husband, so that she can marry again. The 1954 *Black Widow* was 'an electrifying drama about a predatory female', which accounts for the spidery connection in the title. And in 1987 the Black Widow struck again with Theresa Russell playing the woman of whom it was pronounced 'She mates; She kills'.

Horror movies have always been a great place to find venomous animals, and until the advent of the violently awful psychopath genre that first became popular in the seventies, most horror movies were either of the Dracula type, or about the mad professor who carted off nubile young women for a variety of dubious reasons, or about venomous animals. The story in the last type of film is almost always the same. A single specimen or a gang of venomous animals of one sort of another appears to terrorise the inhabitants of a small town until some bright spark thinks of a way of killing them off. Sometimes the animals are of giant proportions, a mutation which is generally due to 'atomic forces'. We have had invasions of giant ants, equally large tarantulas, swarms of bees, rattlesnakes and scorpions; but if you like a good laugh, try and find a marvellously awful film that was made in 1937 called *Sh! The Octopus* about the most brainless detectives you have ever seen fighting a large octopus inside a lighthouse of all places. It is hardly surprising that the motion picture industry has produced so many features that involve horrible creepy crawlies that terrorise the world, but what is interesting is the number of films with venomous connections in the title that have nothing to do with animals at all – the *Snake Pit*, *Spanish Fly*, *Scorpio* and the *Golden Salamander* are typical.

It almost goes without saying that any newspaper reports about venomous animals are sensational and rarely complimentary to the animals concerned. It was therefore surprising to find a recent report in the national press about Botswana headlined 'How Ants Brought Prosperity And Democracy to Botswana'. Botswana is an oddity is southern Africa in that it is comparatively wealthy, and is a genuine democracy. The wealth of the country is based upon diamonds, that strange mineral that has only limited uses but commands such high prices on the markets of the world. Many centuries ago, according to the report, ants seeking scarce sources of water beneath the scorching sands of the Kalahari Desert brought to the surface tiny particles of rock which otherwise would never have been seen. There they lay, nondescript and unnoticed until ten years ago when the huge mining company de Beers began to sift the sands a hundred miles west of the capital, Gaborone, to ascertain whether there might be any mineral wealth beneath the surface. Some of the little particles were found to be kimberlite, the soft volcanic rock in which diamonds are found. Continued exploration convinced the company that it was worth exploring further, and the end result was a mine which has proved to be the richest source of diamonds in the world, a mine which is expected to provide over $1\frac{1}{2}$ tonnes (tons) of diamonds this year alone.

Far more common in newspapers is a headline such as 'Giant Spiders Sent to Property Boss'. The story behind this intriguing title is that a London company director returned from an enjoyable holiday abroad to discover that someone had very kindly inserted three large Chilean Rose Spiders through his letterbox. He found two of them in the hallway of his house on the day he returned, and the third was discovered in the sitting room a day later. They all ended up at London Zoo. The victim of the exercise was asked by police if he had any idea who might have sent him this attractive welcome home present. After some thought the only idea he could come up with was that it could have been a disgruntled former girlfriend. In the circumstances it would seem to have been a remarkably generous and affectionate gesture since the value of these three fascinating animals was about £100, and many tarantula fanciers throughout the country would have been absolutely delighted to receive such a splendid gift. One can understand the poor chap's surprise at finding the animals but he had nothing to fear from them since they are friendly little things that are most reluctant to bite, and even if they do the effect of their venom is no worse than a bee sting. Unpleasant to be sure, but hardly lethal.

One of the nicest newspaper stories was to be found in the *Yellow Advertiser*, a free sheet in north Essex, which described how a former teacher, Kathy Hancock, had just started receiving a government Enterprise Allowance to rear and breed tarantulas, stick insects, millipedes, scorpions and a host of other varieties of many-legged creepy crawlies for sale to collectors. According to the article, Kathy's stock consists of '560 tarantulas, 400 stick insects, 300 giant snails, 15 giant millipedes, 20 scorpions, and sundry cockroaches and praying mantises'. She is quoted as saying that she would like to breed enough of them so that importing – many of them dying on the way or being badly looked after in pet shops – would no longer be necessary. Purely from a conservation point of view Kathy believes the project is worth doing, and this is the attitude of most responsible creepy-crawly owners who are desperately interested in the welfare and conservation of their charges, and thereby, indirectly, of the habitats in which they live.

An intriguing report in a Sunday newspaper sounds as though it might be the result of 'creative journalism' rather than a report of an actual occurrence. It tells of a vicious Blandford Fly which is biting people unfortunate enough to live in Dorset. According to the newspaper the bite of this savage animal causes dizziness, and at the site of the bite a swelling up to 7.6cm (3in) in diameter. Inevitably the fly is said to be rampaging across the county. In this sort of story, insects never drift, swarm or fly; they always rampage.

One music journal in America recounts the story of a folk singer with the delightful name of Rattlesnake Annie. In the best tradition of American folk singers she is said to be half Cherokee Indian. The lady was obviously no slouch when it came to publicity since she recently adopted a rattlesnake at London Zoo. The name of the far better known pop singer Sting could refer to any of a number of venomous beasts, while the pop group Whitesnake is every bit as vague; but though their name does not necessarily have any reference to a venomous snake, publicity pictures show the musicians with paintings of vaguely viperine serpents.

Snake, spider and scorpion motifs are commonplace in contemporary jewellery. A recent visit to establishments in London's Bond Street selling highly priced pieces made from precious metals and flawless gemstones revealed that every single outlet had at least one piece depicting a venomous creature in stock. The famous Greek jeweller Ilias Lalaounis includes amongst his other works, spiders, scorpions, ants and centipedes, and while they are not necessarily to everybody's taste, they are magnificent animals.

The world abounds with venomous motifs. Many large department stores offer cobra ornaments originating from the Far East generally, and India in particular, together with ornamental pieces in a wide variety of forms depicting animals such as scorpions. In shops that specialise in Art Deco, lamps in the form of erect cobras are not uncommon, whilst a gallery in London that deals in designer fabrics for the home stocks a very pleasing rug constructed to look like a piece of grass interspersed with narrow tracks upon which are rows of busy little, three-dimensional ants made from black wool.

In some parts of the world tattoos have become an art form,

and exquisite designs that involve many months of work upon stoical victims can be commonplace. Sadly in Britain the work of most tattooists is pretty grim, and often anatomically incorrect. For years the customers of tattoo parlours entered in order to be covered with 'I Love You Mum' slogans, and hearts and flowers, or skulls and daggers. In the last few years, however, a new craze has emerged for spiders' webs to be

Cobras, spiders and, in this case, scorpions are popular with jewellery designers (John Nichol)

tattooed onto elbows. They look somewhat grotesque, but undoubtedly they are popular.

The world of children is full of tales of venomous animals. Snakes and Ladders must be one of the oldest board games in the world. There cannot be a child in Britain who has not had one at some time. Nowadays the snakes, depicted on the boards are boring old nondescript, all-purpose snakes that could be anything at all, but a look back through some of the splendid examples that were produced during the Victorian era shows that originally the snakes were either the large constrictors or, more commonly, venomous snakes, especially cobras and Russell's Vipers.

We all grow up with Miss Muffet and her arachnophobia, and we all wonder why grown-ups tell us the story of Robert the Bruce and the spider as though it were a great moment in history, and even before we hear of either of these actual personages, we learn the nursery rhyme about Incy Wincy spider. Some children's rhymes go back into history a long way, often emerging centuries later in virtually the same form, but Incy Wincy Spider seems to have started in Victorian times. 'Rikki Tikki Tavi', the delightful story by Rudyard Kipling, is another tale that all of us seem to absorb with our mother's milk. It tells of an eponymous, tame mongoose, the companion of a small English boy in India in the days of the British Raj, that is responsible for saving the lives of the child and his family from the cobras that live in the garden of the house where they live. Since World War II the animals in children's stories have all become bland nonentities with nothing like the character of Nag and Nagini in 'Rikki Tikki Tavi'. They have lost their personality to become Disneyesque, Care Bear bland cuddly toys.

The funny thing is that though venomous animals are usually perceived as evil, occasionally one comes across a character that is presented in a totally different light such as the spider Charlotte in E. B. White's *Charlotte's Web*, and Money Spiders are universally regarded as being lucky. Unfortunately, most people still seem to have no compunction about killing spiders, even though they are almost all harmless and are good to have around the house to get rid of the multitude of noxious pests that would otherwise abound. On the other hand some people, whilst not wanting them in the home, are reluctant to kill any they may find there. This attitude has led to the manufacture of items to get rid of the little animals. One is a remarkable device rather like a pair of scissors except that one of the blades is a disc and the other is a dome that fits over it exactly. The idea is that you place the flat part in such a position that the spider walks onto it, then you slide the dome over by closing the handles of the scissors so that it completely covers and traps the spider beneath, after which you can carry it outside and dispose of it humanely in the garden. Another device on the market is designed to solve the eternally distressing problem of the spider in the bath. Clearly you want to fill the bath with hot water, but to do that would kill the spider that occupies it. So what you want to do is go out and buy a spider ladder which you suspend from one of the taps. The spider can then climb up to safety!

(right) *The handsome King Cobra is the largest venomous snake*

(pp154–5) *The Viper family causes many fatalities. The main picture is of a European Viper (or Adder) and the inset picture shows a Pit Viper being milked of its venom. The venom is collected to produce a treatment for their bites*

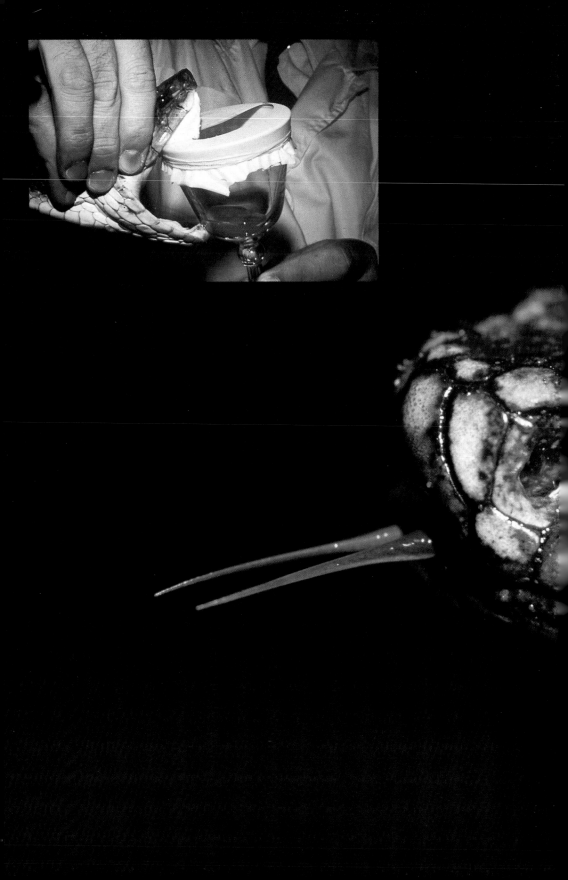

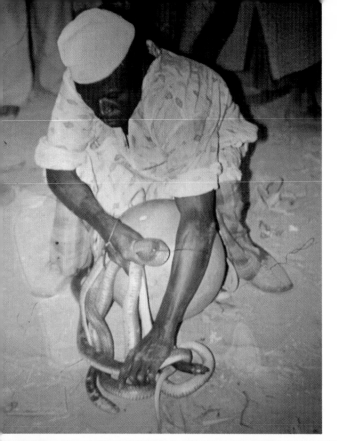

Snakes cannot hear, and their dance is merely an attempt to keep an eye on the end of the moving flute. Though snake charmers are traditionally considered to be Indian this one is in Kano, Nigeria

The European Viper (or Adder) is the most northerly of all venomous snakes

8 · The Keeping of Venomous Animals

On the face of it, the idea of keeping venomous animals in a private house might seem quite absurd, but whereas it is perfectly ridiculous to want a rattlesnake in a tank next to the television in the sitting room, it should be remembered that the vast majority of venomous animals are relatively harmless, either because their venom is of insuffcent virulence to do very much damage, or because they are so placid that they do not hurt anyone. Neighbours are frequently the biggest obstacle to keeping such species, and one could be forgiven for believing that such neighbours are more a problem than any animal could ever be. People have been keeping bees for centuries and nobody raises an eyebrow. This general phobia about venomous animals of all sorts is something that you have to put up with if you are interested in these creatures; it is more or less universal.

Unfortunately the world is full of people who wish to keep venomous animals because they feel it gives them a macho appeal, or perhaps so that they can frighten their friends with their charges, or impress them by handling these terrifying denizens of the jungle, which is invariably how such people refer to their animals. They have given the whole business a bad name and it is hardly surprising that they reinforce the prejudice that folk already possess. Because this irresponsible attitude exists, a few years ago in Britain legislation was brought in to restrict the keeping of wild animals. It is known as the Dangerous Wild Animals Act, and although it is a great idea in theory it often turns out to be quite ridiculous in practice. What it says is that no person may keep an animal that is officially classed as a dangerous wild animal unless they have a licence from the local authority to do so. These licenses are fairly expensive to obtain, and before one will be issued the inspecting officer of the authority has to ensure that the applicant is a responsible person who is going to house the animal in a secure and reasonable fashion.

In many parts of the country the system works well. In others it

Why Spider Plants are called such is beyond comprehension, since they do not look like spiders, not do spiders especially associate with them (Keith Stiff)

is a farce. Inspectors are often people who have no idea what they should be looking for in someone who has applied for a licence, and would not know proper housing for the relevant animal if they fell over it. Consequently some people are being granted or refused licences for all the wrong reasons. This is not necessarily an insuperable problem since, although an animal dealer is not supposed to sell any animal on the schedule of the Act to anyone without a licence, there are always shady characters who will do so if the price is right. If such an animal then produces offspring, someone who has obtained it illegally in the first place is likely to sell the babies in the same manner.

The other silly part of the whole business is the list of animals that are classed as dangerous. No one would argue that rattlesnakes or tigers should be labelled in this fashion, but some of the listing is quite ridiculous. The Pigmy Marmoset, a tiny monkey-like animal from South America is one such creature. The little beast would easily fit inside a teacup and still leave room for milk and sugar. It is not venomous, and though it can bite it is not nearly as dangerous as a cat or a dog. Though the government claims to take advice from experts before any animal is added to the list, one cannot help wondering at times where their expertise lies. There are several reasons why Pigmy Marmosets should not be kept in captivity without a licence, not the least being that they have highly specialised requirements and ought not to be undertaken by people who do not know what they are doing, but to class them as Dangerous Wild Animals is perfectly ludicrous.

Not many years ago anyone could keep anything, which meant that it was not uncommon to find tanks full of venomous creatures in the oddest places – at least one man had a large tank full of very dangerous venomous animals in the sitting room of his council house. Tanks often had no security locks and children could easily have lifted the lid. The astonishing thing is that more accidents did not occur, which shows what nice people animals are! With the advent of the Act this sort of thing was stopped and nowadays it is virtually impossible for the wrong people to get hold of venomous animals, though it does happen occasionally. Caring, responsible, knowledgeable people should not have too many problems obtaining a licence, and although some organisations object strongly to the keeping of any animals in captivity, the activity has provided a wealth of information about animals that would not otherwise have been obtained, and is one way of helping to conserve species that are struggling to survive in a fast disappearing wild.

Legislation similar to the British Dangerous Wild Animals Act is found in the USA and various other countries, which means that the range of venomous animals one can keep without a licence is somewhat limited. Specialist keepers of venomous animals will not need basic information on how to care for their charges since without it they are not going to get their licence anyway, but for those who find the whole topic fascinating, it may be useful to have a few pointers. Nor should such persons be put off because an animal is venomous, for those that are available without a licence are not too dangerous, though care should always be taken with any venomous animal, however mild the effect of its venom. Keepers have a responsibility to the animal and a responsibility to other human beings, and even though one person may not suffer from a venomous bite, another may be allergic to something in that venom and be affected badly. If an accident occurs it will be your fault; you cannot blame the animal. But a bite may cause considerable pain and distress to a friend of yours as well as calls from the public that the animal be killed.

There are also one or two species, not yet placed upon the schedule of the Dangerous Wild Animals Act, that can be very dangerous indeed, so do not even think of keeping venomous animals unless you are interested in studying them and breeding from them and, if you do, do it responsibly. This attitude extends to the keeping of comprehensive records of the animals' care and behaviour, and taking photographs at each stage of the animals' lives. Not only is this sort of record-keeping absolutely vital, it adds considerable interest to the whole business. There can be no enjoyment for anyone in simply throwing a lump of food to an animal each day and forgetting about it for the rest of the time.

The other thing you should do if you are contemplating taking up the keeping of any animal is to join a responsible society. Such bodies take considerable care to show the outside world that their members are not peculiar because they care for snakes or spiders. They are also an essential source of information and many of the members are extremely knowledgeable people. The following list of national organisations should be useful. They will be able to put you in touch with local groups and other like-minded individuals in your area:

This man was bitten by a pit viper in Costa Rica. Damage is so great that most soft tissue below the knee has fallen away to expose the bones (Herbert Lieske)

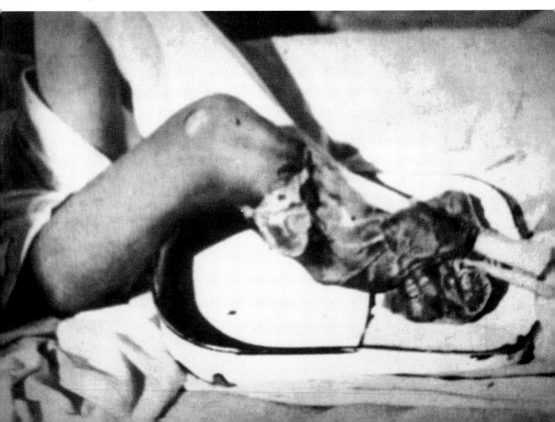

British Beekeeping Association: National Beekeeping Centre, National Agricultural Centre, Stoneleigh, Kenilworth, Warwickshire. OV8 2LZ.

British Herpetological Society: c/o The Zoological Society of London, Regent's Park, London, NW1 4RY.

British Tarantula Society: 36 Phillimore Place, Radlett, Herts. WD7 8NL.

International Herpetological Society: 65 Broadstone Avenue, Walsall, West Midlands, WS3 1JA.

The American Arachnological Society: c/o The American Museum of Natural History, Central Park West, 79th Street, New York, NY10024, USA.

The Chicago Herpetological Society, 2001 North Clark Street, Chicago, Illinois 60614, USA.

The Long Island Herpetological Society, 117 Santa Barbara Road, Lindenhurst, New York, NY11757, USA.

The Society for the Study of Amphibians and Reptiles, c/o The Department of Zoology, Miami University, Oxford, Ohio 45056, USA.

A most useful publication, *Herpetological Societies Worldwide*, is available from Fauna Classifieds, 1035 Middlefield Avenue, Stockton, California 95204, USA, who can also supply several other valuable publications to anyone interested in herpetofauna. Both in Britain and in the United States, any aquarists' shop will give you the address of your local group.

Each organisation produces its own magazines and other literature, but in addition the following journals are worth getting:

The Aquarist and Pondkeeper, The Buckley Press Ltd, 58 Fleet Street, London, EC4Y 1JU.

British Bee Journal, 46 Queen Street, Geddington, Kettering, Northamptonshire, NN14 1AZ.

Snake Keeper, 159 Stanley Hill, Amersham, Buckinghamshire, PH7 9EY.

Today's Aquarium, Aquadocumenta London, 70 Wood Vale, London N10 3DN.

Tropical Fish Hobbyist, TFH Publications, Jersey City, New Jersey, USA.

The Vivarium, The American Federation of Herpetoculturists, P.O. Box 1131, Lakeside, California 92040, USA.

YES Quarterly, c/o Department of Entomology, College of Agricultural Sciences, Texas 79409, USA.

The next essential is to read as much as you can about your subject; the Bibliography lists several useful volumes.

The reason that venomous animals are venomous, is to a great extent to enable them to catch their prey. Someone who keeps a pet budgie needs a cage with a budgie in it, and a jar full of seed and grit. Someone who keeps venomous animals on the other hand has to be aware that he is going to have to supply suitable prey for his charges. Sometimes this will be live animals, and sometimes freshly killed ones, but either way breeding colonies of species suitable as food can take up a fair bit of space. The alternative is to buy your food from a dealer, and if space and time are at a premium you may have to do this, but it is far cheaper to breed your own if you can. An intricate operation in itself, this is not the place to discuss it in detail, but Ann and Frank Webb's *Breeding Live Food For Reptiles and Tarantulas*, and the present author's *The Complete Guide to Pet Care* are two books that will explain everything. The trouble with keeping something as fascinating as venomous animals is that your collection soon starts to grow and expand, so whereas the only food you had to worry about at the beginning might have been mealworms, before long you may be knee deep in maggots, stick insects, locusts, crickets, mice, cockroaches and chicks.

HOUSING

Secure housing is absolutely essential whatever animal you are keeping, and the best of all for any venomous species is an aquarium tank with a lid that you can close and lock. For the larger species of snakes, an old glass shop counter is perfect but they are as rare as hens' teeth these days, and on the odd occasion they can be found they cost the earth; but keep an eye out for any local shops that may be closing down and you may be lucky. It should be realised that almost all the animals you may want to keep need some form of heating. It is true that with the advent of central heating many of them can be kept nowadays without extra warmth, but if you do rely on central heating you must make sure that the room in which the animals live does not get too cold during the night. If you intend to heat your tanks individually there are various ways of doing this and reading appropriate books will present you with your options.

The siting of a tank is important. If it is to house venomous animals, however innocuous they might seem, it must be in a special room where unauthorised visitors and small children, whomever they belong to, cannot go. The room should be escape-proof as all snakes and invertebrates are Houdinis and will get out through unbelievably small gaps and, while we're on this subject, do not make the mistake of placing a sheet of Perspex or something similar over the top of your tank while you go next door for some clean water. Snakes, and even large spiders, are quite astonishingly strong and can very often lift a weight that you might think beyond them. The tank should be as large as possible. That does not mean it is necessary to provide a 3m (9ft) tank for a tarantula. If you do, you will soon discover that the spider lives in quite a small area and will not use most of his enormous domain. One needs to be practical, and a spider can make do with a surprisingly small home, as will many invertebrates, though orb web spiders do need quite a large area, so that they can build their beautiful webs. At least one person I know suspends an old hula hoop vertically in the corner of the room and his orb web spider happily uses this as a frame for the web.

When you come to arrange the inside of the tank for your animal you should bear in mind the conditions that prevail wherever the animal comes from originally. That does not mean that they should be slavishly copied, and if you try and set up a tank with humidity at almost saturation point for some rainforest species, you will find that you are forever fighting disease of one sort or another as the little micro-organisms causing it reproduce most efficiently in these conditions. There is constant discussion as to what makes the best substrate for terrestial species. Peat works fine for some species, and gravel for others. Sand is dreadful stuff and should be avoided at all costs, though somewhere you can be sure somebody is using it and swearing by it – this is the nature of animal keeping. Vermiculite is gaining popularity as a substrate, for several reasons. It does not encourage bacteria and fungi, and can be cleaned when necessary. It retains moisture if you are trying to reproduce a humid environment, and it will stay dry if what you want is a desert, and it does not look too unnatural. At least one herpetologist uses paper kitchen towels on the floors of frog tanks, and though she swears that they are the best thing ever, they look absolutely awful. Some animals like to burrow, and Vermiculite suits this behaviour as well.

The tank should also be fitted with some sort of hiding place for whatever is going to live in it. A flowerpot turned on its side is popular and frequently used, though it looks somewhat artificial. The trouble with using a couple of rocks to support a piece of slate or something similar is the ever present danger that the roof might collapse and squash the inhabitant. Animals are far stronger than many give them credit for and will frequently shift a rock that looks far too heavy for them. If you intend to use an overhead light, and there are good reasons for not doing so even though they are traditional, there should be part of the tank where the animal can get away from the heat source when it wants to. And never put the tank in a window, as sunlight can raise the temperature alarmingly and cook the inhabitants. Nor should an ordinary household light be placed against the side of the cage to raise the temperature as an emergency measure for whatever reason. It is the easiest thing in the world to leave it for too long. After all, watching the mercury creep up a thermometer is hardly a riveting pastime, and one tends to guess how long it is going to take while some other task is performed. The result is inevitable. You forget, only to remember later in a panic and rush to the tank to find a cooked snake.

An often overlooked aspect of animal husbandry is the importance of humidity. The correct level of atmospheric humidity is absolutely vital but it is not always easy to get right, for if it is not high enough the animal may suffer from dehydration or have trouble with normal life processes like sloughing and if you get it too high the result is that fungus and disease begin to grow in your tank with all sorts of nasty results.

Plants in a tank look lovely, but unless the enclosure is enormous they usually turn out to be a complete waste of time. They tend not to survive very well, though you would think they would in a mini greenhouse, and animals play havoc with them. A heavy snake weaving through leaves is certain to break them and animals with tiny claws usually ruin them quickly. Often the best bet is to decorate a cage with plastic plants, an effect which rarely looks very good, but at least it is not bare and stark, and the inhabitants can hide in the foliage or go for a climb now and again. Beyond that, all that is needed is a water dish, secure or heavy enough not to tip over, and of a nature such that an animal can escape easily, and not drown.

Venomous aquarium fish should have an aquarium according to normal fish-keeping requirements, but the basic living requirements of terrestrial animals (ie space, warmth, light, food etc) are just as valid for fish.

One important thing to remember if you are keeping invertebrates is never to use an insecticide in the house or you may wipe out all your stock.

BUYING AND CHOOSING

You are now ready to go out and buy your animal. Experience is the best asset when it comes to choosing livestock, and if you haven't got it the next best thing is a knowledgeable friend. The biggest threat to the sort of animals you will be looking at is dehydration. If you can possibly avoid it do not buy your first two or three animals from a pet shop. There are some perfectly good pet shops, but the more exotic animal forms do very badly in most of them. It seems crazy for a pet-shop owner not to look after his stock even if he is not interested in their welfare, since spiders, snakes and so on are usually fairly expensive animals and he cannot afford to lose many of them, yet frequently pet-shop reptiles and invertebrates are kept in very poor conditions. These unfortunate animals may also be diseased or covered in parasites or have lumps missing, but dehydration, as we have seen, is what you have to watch for above all else. It is difficult to describe what dehydration looks like, which is where your friend comes in. All one can say is that a dehydrated animal looks dull and lifeless, even when it is moving around. If the cage is in a state as well, the chances are that the stock is not cared for. There should be a pot of water present that the animal can reach at all times. That might seem obvious, but not all pet-shop cages do have water and of those that do, not all can be reached by the animal for sale.

By and large, it is infinitely better to get your animals from a specialist dealer or from another member of one of the clubs you have just joined. Either way you are almost certainly going to pay less for your animal than if you buy it in a shop.

Most of the venomous animals that are commonly kept and which do not require a Dangerous Wild Animal licence, are invertebrates. Spiders of many species, mostly theraphosids or bird-eating spiders – or, if you must, 'tarantulas' – are imported regularly by specialist dealers. A few species of other spiders are normally available in small numbers, such as various orb web spiders; and the occasional Black Widow or real tarantula – true tarantula, it is usually called – figures on dealers' lists. Scorpions too are regularly available for sale, and a few other odds and ends, such as giant centipedes and some animals that, while technically venomous, are no threat to humans. These are things like Giant Millipedes, Whip Scorpions, and some caterpillars which have urticating hairs along their bodies.

It is true that some species of back-fanged snakes may still be imported freely. They are not commonly found on dealers' lists as they are usually specialist feeders which means that they are difficult to sell. Long-nosed Whip Snakes come in sometimes, not usually because the importer has ordered them but because they have been sent in place of something else which he specially requested. These delightful little animals are very common in the wild, so unscrupulous dealers often send them instead of more desirable species. The great problem with them is that they are usually badly dehydrated when they arrive and their long, very delicate noses are usually damaged from being rubbed on coarse cloth bags, or on fine wire mesh in the holding cages of the Far East prior to despatch.

If you see a snake in such condition, leave it be however sorry you feel for it. Establishing it in captivity will be almost impossible. It is difficult enough with healthy, newly imported specimens.

Sometimes Flying Snakes arrive in this country, or else their close relatives, Paradise Snakes. They are beautiful little things, particularly the former, but as with the Long-nosed Tree Snake the great difficulty is to wean them onto a suitable diet for captivity. In the wild all these snakes feed on lizards, which goes against the grain somehow in captivity, and apart from anything else lizards are expensive to buy if you have to feed a knot of writhing tree snakes. In a case like that you have to tax your ingenuity to the extreme until you finally persuade the reptiles to take something like mice. It doesn't always work, and many snakes would rather die in captivity than take an unfamilar food, in which event the only and most unsatisfactory option is to force feed them regularly. This is a traumatic business for a snake, and should never be undertaken by an inexperienced person.

If one is to keep carnivorous invertebrates successfully, feeding them livefood is usually essential, but you have to take care with some forms. If the food animal is not removed in a day or so, assuming your specimen is not hungry, real problems can be caused on occasion by the carnivore attacking and eating bits of it. It should be noted that it is actually illegal to feed live mice and chicks to snakes, lizards, amphibians and spiders.

A specialist in King Cobras told recently of the job he had trying to persuade one of his animals, which was a snake eater, to settle for something less exotic and expensive in captivity. For the first few months it would only take snakes. Later a small piece of raw beef was tied to the tail of the dead snake that was being fed to the King Cobra. There was no problem with this, so over the next few weeks the piece of beef got longer and longer and longer without the snake appearing to notice. Finally it was content to eat long, thin snakes of beef on their own. Muscle tissue on its own is not a complete diet, so the next stage was to tie a small dead mouse to the tail of the beef snake. In due course the mice got bigger, and were eventually changed for rats. A final stage in this long-running saga was to reduce the size of the piece of beef until the cobra was taking only rats. It might sound like a gruesome comic opera, but if you want keep reptile-eating snakes, these are the lengths you have to go to.

Other venomous reptiles are rarely to be found on the market these days. Until recently it was possible to buy, quite legally, specimens of the glorious black and gold Mangrove Snake but that now requires a licence. It is interesting that that particular species has been put on the Dangerous Wild Animals list, because although it is undoubtedly venomous, there doesn't seem to be a single record of this snake biting anyone. Thus there is no way of assessing how much damage would be done to a human by a bite from one of them. Like most back-fanged colubrids Mangrove snakes can grow to be large animals and it is wise not to take liberties with them, though the chances are that the effects of injected venom would be slight. As stressed before, nothing is ever simple in the world of venomous animals and it is interesting to note that John Coborn, in his book *Snakes and Lizards*, describes Mangrove Snakes as irritable, as this conflicts with reports by other experienced herpetologists. But everyone can tell you of contradictions like this. One snake keeper mentioned how irascible he found Boomslangs, while another commented on how placid they were.

BREEDING

Whatever animal is of interest to you there is no longer any reason these days for keeping single specimens. Animals the world over are under threat and it is essential that collectors of any species breed them if they possibly can. Many responsible keepers do this already and the various organisations devoted to the keeping of particular types maintain stud books. Each member makes reports whenever he obtains new stock and these are added to the master file so that whenever a member finds himself with a mature animal, but does not have a mate for it, an appropriate animal of the opposite sex can soon be located in order that the two can be mated. Some sort of arrangement is then entered into regarding the offspring. They are usually either divided between the owners of the two animals, or each takes the few he wants and the rest are sold off by the society which shares the money between the two owners and keeps a commission which goes into society funds.

Spiders present a particular problem with regard to breeding. A mature male lives for a very short time, which means that once he has become mature, arrangements have to be made to get him to as many suitable females as possible. Furthermore the situation is aggravated because male spiders are often not as large or as showy as the females, consequently they are not imported as commonly as the opposite sex. As if all this did not present enough difficulties, the whole business of mating spiders is fraught with danger. It is well known that female spiders tend to attack males. Male spiders know this just as well as you do, so the whole procedure is long and carefully conducted, and it is pointless trying to hurry any of it. As soon as two suitable spiders are put together there follows a careful approach by the male that may take a very long time. If the female is co-operative the male calms her with ritual stroking and other pre-copulatory activities. When he feels the time is right he will move nearer, taking care all the time and remaining ever watchful until he is able to mate. Afterwards he runs like hell, and has to be removed to safety immediately. If all has gone well, the female will remain still for a few moments while the male has a chance to get out of the way. Spider fanciers tend to have perpetual worried frowns!

Spiderlings are delightful little creatures, and probably one of the most appealing is the baby of the common Pink-toed Tarantula. The adults are black all over except for bright pink fur on their feet. The babies on the other hand are a very pale pink apart from black boots which look most attractive and fashionable. Great care has to be taken with spiderlings. They are tiny, and born escape artists, and are the easiest things in the world to overlook if they get out of their tank. Each has to be kept individually, as they are not averse to eating their siblings, and in the early stages most people keep them in small plastic containers with transparent lids. Feeding time can be quite a problem as no self-respecting spider has one or two babies. She has heaps, and there must be occasions when everyone who keeps these animals must wish he owned a budgie instead. Inserting small crickets one by one into dozens of small plastic pots without spiders dashing out, and without crickets escaping all over the house, can be a time-consuming exercise.

Such mayhem does not last long, however. There is always a ready demand for captive-bred spiderlings of any species.

9 · Conservation

Venomous animals of all sorts invariably receive a bad press, and it is an uphill task persuading anyone that there is any point at all to saving them. But it is every bit as important to conserve our venomous animals as it is to save the world's cuddly animals such as pandas. Tell someone that you would rather be bitten by a tarantula than have a close encounter with a Giant Panda, and he will not believe you. The reasons for conserving pandas and spiders are the same though. If we destroy or introduce a species, deliberately or inadvertently, we have no idea what ripple effect that is going to have on the environment. It has been shown over and over again that the loss or introduction of a single species can cause untold unforeseen problems.

A typical case was the introduction into some West Indian islands some years ago of mongooses from India to control the venomous snake populations in the sugarcane plantations. The thinking behind this crazy idea was that in India snakes and mongooses are mortal enemies, we all grow up being told that, therefore these fierce little predators should sort out the Fer de Lances and coral snakes in among the sugarcane without any difficulty. The newly arrived mongooses wandered off and ate a few snakes, and then one of them discovered that the plantation workers kept little, scrawny chickens. No Caribbean chicken is going to put up the fight that a Fer de Lance is, and there is infinitely less risk to a mongoose from a chicken. Word got around and before long chickens were disappearing fast. When they had gone the mongooses started on anything else of a suitable size and soon there was not much left in the way of wildlife apart from snakes. With no other natural controls the latter had a population explosion, and soon the fat mongooses were sitting around in the sunshine of this land of plenty while the islands had more venomous snakes than ever. We can never tell what is going to happen to an environment when we start to change it.

One aspect that should not be overlooked is the active role that venomous snakes play in many primitive agricultural areas of the world. Cobras and other snakes claim many tons of rodent pests each year, and in areas where venomous snakes have been extensively collected for a number of years, local peasants will tell you that grain yields have decreased due to pest despoliation.

Another good reason for conserving stocks of all animals is that people from other countries are happy to pay good money to come and see the fauna of the host country. East Africa was one of the first parts of the world to actually capitalise on this and nowadays there is a considerable industry involved in the promoting of safari holidays. Tourists bring in large quantities of valuable foreign exchange, and many people are employed by the hotels and safari companies as guides and to man soft drink stalls and to service all these visitors. It is easy to point out that Americans go to Kenya to see lions, not snakes, and of course this is true, but in Florida there are many tourists who visit parts of the state just to see alligators, or to listen to the enormous congregations of frogs in some of the swamps. Snakes or spiders may not be to everybody's taste, but each year more and more people around the world are keeping these animals, and membership of herpetological and arachnological societies is increasing at a phenomenal rate. All these members are happy to go on snake-watching or spider-spotting holidays, while aquarists who keep venomous fish are happy to spend hundreds of dollars snorkelling in the Red Sea to watch Lion Fish.

A third excellent reason for saving the world's venomous animals is that some constituents of their venom are used in the treatment of various human ailments. We need a continued supply of venom for this reason, and who knows what is still to be discovered that might be of use to man. Once species are destroyed we will never know. The Madagascar Periwinkle, a small plant with pink flowers, is now extinct in the wild, destroyed with all the other species in certain areas to make room for development. Luckily there is a plentiful supply of this plant in cultivation, from stock that some far-seeing person rescued before its total destruction. Luckily, because it is now the source of a drug that is used successfully in the treatment of leukaemia in children, many of whom would have died without it.

Then there are the aesthetic and moral objections to destroying animals. They are just as valid as economic ones, but sadly today when there is a straight choice between money and conservation, money is invariably the winner. The biggest threat to any animal is not the destruction of that particular species for its skin or for its meat, or for that matter its

The flowers of spider orchids are said to resemble spiders (Peter Wakely/ Nature Conservancy Council)

venom, but the removal of its habitat. Destruction of habitat is a term that in practice has lost any significance, but rainforest, wetlands and other natural habitats are disappearing at a phenomenal rate. A country like Thailand, for example, has lost something like four-fifths of its rainforest since the end of World War II. It is fashionable to blame peasants in the Third World for chopping down trees for firewood and practising slash and burn agriculture, but this is only a way of sweeping the problem under the carpet. The truth of the matter is that places such as tropical forests are disappearing because of consumer demand. Look around you, where you are sitting now, and it is almost guaranteed that within your field of vision there will be at least some tropical hardwood. There is almost no doubt that it did not come from a plantation. Five years ago, or fifty years ago, that door sill or coffee table or window frame was a tree growing in Amazonia or south-east Asia. It was a tree full of insects, and nesting birds, and small mammals. As it fell, it destroyed many of these life forms, and crushed other trees on the way down. Finally it was dragged out to the roadway, flattening yet more plants and animals on the way. Any wildlife left flew away only to find that the adjacent bits of jungle were also falling around it.

In the last few years the rate of destruction has increased alarmingly, and whereas previously most of the work was performed laboriously with unsophisticated hand tools, nowadays power-saws and huge vehicles can destroy an acre in the time it took to cut down a single tree not many years ago. Much of this disappearing habitat is already protected, but when the price being paid for the timber is high enough, it is very easy to buy off forest guards and junior bureaucrats, and even government ministers. It is easy to put the blame onto them, but while demand exists from the Western world it will continue to happen. Every single journal on the bookstands that covers the home and garden is packed full of features and advertisements that positively advocate the use of tropical hardwoods. Mahogany-framed conservatories are weatherproof, teak furniture is all the rage, and fittings and ornaments for the home from ebony and purpleheart and ironwood are all around us. One or two firms are finally looking at other options but the inertia that has to be overcome is unbelievable. Some timbers have been taken for granted for so long that we think of them in the same way as we think of pine, from renewable resources; but our broom handles are made from ramin, a tropical rainforest tree from the Far East which has been exploited in such quantities that Indonesia has finally banned its export. Things are beginning to change because they must. Would it not have been better if people had listened years ago when warnings were first sounded?

It is not only the demand for timber that is responsible for the loss of valuable resources. The buzz word these days is fast food. This means the increased production of hamburgers and the like. Hamburger meat does not come from a packet; it comes from a cow, or at any rate a bullock, and to meet the increased demand enormous ranches are now being built in the most unlikely countries to rear the animals to provide the beef. One cowful of hamburger meat needs an awful lot of grass, and the acreages of these places are enormous, all hacked from natural landscapes especially to increase the dividends for the shareholders of the fast food companies.

So, what to do about it all? Well, to start with it is important that each person makes a conscious decision not to partake in the causes of the destruction. There is no point in beating one's breast over what is already done, but demand can be halted. The next thing to do is to educate friends. It is no longer considered cranky or weird to express an interest in the environment, and when enough people are aware of the problem they can put pressure where it is needed. One would think there could not be a single person these days who is not conscious of what is happening, but the extraordinary thing is that most people either do not know, or do not care. Having said that, one can understand an unthinking person turning the page of his newspaper or switching channels on television when environmental issues are raised. After all, unless one is committed, the whole subject is hardly inviting. It is always presented in such an unappealing manner. It seems to be a universal law that whenever someone cares desperately about any topic at all they lose their sense of humour and can develop an instant ability to cause terminal boredom in anyone to whom they speak. You've noticed? Great, then make sure you don't join them. It is always difficult to try and persuade people to your way of thinking as they tend to dig in their heels. Just tell them, and leave it to them.

The people who are most difficult to understand are those who go on holiday to some exotic location, and on the day of departure look frantically through the airport shops for souvenirs to take home. Some bring back picture postcards, which is fine. Some bring back pottery guaranteed to have been made by local craftsmen, which is usually a bit kitsch and invariably overpriced at airport shops, and some bring back stuffed animals that have been fashioned into the most grotesque artefacts. Some of those departing tourists think briefly about the legality of importing such articles into their homelands, and are pleasantly reassured by the smiling sales staff that there will be no problem. They become quite shocked and often very angry when Customs officials confiscate the things on their return. Whatever someone in a shop might say, most wildlife is protected these days and it is an offence to import it into another country without appropriate documentation. The Customs and Excise departments of most Western countries hold stocks of the oddest items. Ivory, crocodile-skin handbags and furs of spotted cats are the commonest articles, but there are some real peculiarities. Stuffed cobras are two a penny, some of them wrapped around balding mongooses. Stuffed frog orchestras, each overfilled amphibian fitted with glass marbles as eyes, are pretty revolting; but the hoof and foot of a zebra, cut off at the knee so that a cigarette lighter can be fitted at that point, is one of the most bizarre of all.

Education is vital. If only Customs departments had the money to take exhibitions of this sort of material to every school in the country the message would soon get through. It seems that each succeeding generation of schoolchildren is more environmentally aware than the last and thirsty for information which is often lacking. Which brings one to the last thing that can be done though it is not for everyone, namely the maintaining of captive breeding programmes of threatened animals. Those collectors who keep venomous animals have an especially important role to play, for not only are they enabling species to survive when, despite conservation legislation, their wild populations are being wiped out in the countries in which they live, but they are in a superb position to educate the layman into realising that simply because an animal is venomous, it is not necessarily going to rush about killing every human being it sees. Take some of these animals to a school and talk to the children, and at the end of an hour one can see that they are never again going to shudder when they see a snake, nor squash the next spider that they find in the bath.

Apart from all that, many creepy crawlies have not been extensively studied in the wild. The large, the furry and the valuable have been prodded and poked and investigated *ad nauseam*, while scorpions and centipedes have been largely ignored. Breeders of these animals have a vast store of useful knowledge in their hands that is not available in academic circles. Not only is this information intrinsically interesting, it is going to become even more important in the future as more and more species vanish forever.

10 · Rattlesnake Roundups and Killer Bees

Animals throughout the world are fighting all sorts of pressures to survive. In the case of venomous animals the pressures are perhaps highest of all. Some are common to all animals – the destruction of their environment, the killing of them for food or to supply the fashion and souvenir trade, and their depletion as a result of pollution. In addition to all these factors venomous animals have to put up with one other, the deliberate and gratuitous killing of any venomous animal by man, simply because it is venomous.

In most parts of the world where there are appreciable numbers of venomous beasts, it is perfectly true to say that they are killed when they are found in the open, partly as a precaution, and partly in an act of awful revenge for crimes committed against humanity in the past.

Not far from Bombay in India in the 1800s there was a bounty on Russell's Vipers, and an average of 225,000 Rattlesnakes were brought in every year. The record was a staggering 115,921 specimens within eight days. Despite this sort of depredation and increases in demand for agricultural land and other pressures on the snakes of that country, hundreds of thousands of this particular species are killed and exported, both illegally and legally each year.

An even more extreme folk hatred of snakes exists in the United States, where appalling destruction of rattlesnakes takes place. In some parts of the country the local people hold rattler roundups, which become the centre and focus for fairs and public holidays when folks from miles around collect rattlers in enormous numbers. These are usually kept in pits or large containers such as oil-drums until the culmination of hysterical activity which exacts a devastating toll on the rattlesnakes of the region, and the numerous other animals with which they share their environment, because the collectors are not content with catching single specimens as they find them, instead they dig out hibernacula and pour petrol down holes that are suspected of housing rattlesnakes, and pump car-exhaust fumes down others.

As a direct result rattlesnake numbers are declining fast, and in some states populations are only a tiny fraction of what they were not many years ago. There is in the US an organisation called the International Association of Rattlesnake Hunters. Each year they converge on the small town of Okeene in the state of Oklahoma for their rattler round-up. What started as a means of rattlesnake control (which had a handy spin-off in that it brought in money for local inhabitants who sold snakes to laboratories that milked them for their venom) has degenerated into an annual orgy of destruction.

It is still true that numbers of the snakes are sold for the production of antivenom, while others are used in the preparation of canned rattlesnake meat and in the manufacture of snakeskin products for the tourists who flock to the town. But the whole emphasis of the event has changed to become a major social occasion in that part of the country. For weeks before the roundups, caravans and cars arrive from all parts of the south-eastern part of the country, and indeed from farther afield, to take part, and well before the first day all accommodation in the town is filled. The sole purpose of these roundups is to catch and kill rattlesnakes, even though many of the events during the celebrations claim to have other aims. Members of the association, and anyone else who fancies his chances, go out to catch what rattlers they can, and ere long the pits and other containers in which the snakes are kept, begin to fill up.

While all this is going on, much else is happening. There are competitions to find Miss Rattlesnake of the year, and displays of marching bands and majorettes, while hosts of side-shows and stalls are set up for business. Much rubbish is bought and sold, most of it with some sort of rattlesnake motif, and everybody has a rare old time frightening themselves silly with tales of rattlesnakes that are bigger and fiercer and more dangerous than in former years. One can hardly blame the inhabitants of the area for cashing in on the annual influx of onlookers who are only too happy to spend money in their hotels, and in buying enough local beer to float New York.

The rattlesnake hunters take it all very seriously and there are many competitions to find the heaviest rattlesnake and the longest, and awards are presented to the hunter who manages to catch the greatest number of the reptiles. Each year over a ton of rattlesnakes is collected during this event and the enjoyment of the human participants is assured when they watch the way the animals are handled after capture. There is always a large and appreciative audience for the ceremonies of rattlesnake decapitation, when wriggling snakes are hauled from their containers and their heads are placed carefully on chopping blocks. A swift stroke of a machete severs the head from the body, which is then often cooked so that participants can enjoy eating rattlesnake, perhaps for the first time in their lives. At some stalls, diners are given a certificate to state that the holder of the document has actually eaten rattlesnake meat.

Many other tasteless, gruesome events are enjoyed just as much by the public, and hunters who have reputations for catching numbers of the snakes are regarded as heroes by the sort of people who go to these things. Watching rattlesnake roundups and seeing the atmosphere they generate, it becomes plain that there is more happening than a simple rattlesnake hunt. Those who attend obviously derive as much enjoyment from watching the humiliation and death of large numbers of rattlesnakes as was enjoyed centuries ago by the people who watched bear baiting or cock fighting. Venomous animals in general, and in North America the rattlesnake in particular, have a special place in each person's sub-conscious. No animal in that country causes such revulsion and hatred as the rattlesnake; feelings that go far beyond the damage that these animals actually cause. We have already seen that more people in the Unites States die each year as a result of stings from wasps and bees than from the bites of venomous snakes of all sorts.

Sweetwater, in Texas, is another renowned centre for rattlesnake roundups, and there are about another thirty in various southern states, but around two-thirds are in Texas and Florida. Although the local peo-ple in each community can make money from servicing the visitors who attend these roundups, at least one person in Sweetwater was heard to complain bitterly that the hunters were a far greater nuisance than the rattlesnakes ever were, and this seems to be a common opinion among the occupants of the towns unfortunate enough to be chosen as hosts.

The events are usually arranged for when the weather is mild, and since snakes tend to be most active in spring, even in those states where the winter is not too cold, this is the season when most of the roundups are held. Even the dedicated rattlesnake hunters who continue to deplete the wild stocks to the best of their ability bemoan the fact that the numbers of rattlers are declining. In Okeene just after World War II, colossal numbers of snakes were collected. In 1954 this had decreased to a mere 1,576, and recently the total catch was down to fewer than 500. Tales of declining numbers are exchanged during the evenings when participants sit around the rattlesnake barbecues, for the telling of stories is an important part of these ceremonies. Inevitably the best stories are related by members of the White Fang Club. The sole qualification for entry to this elite is to have been bitten by a rattlesnake at some time.

Later in the evening as the beer flows more freely the stories become taller until sometimes they bear little resemblance to reality, and the ones that got away get longer and longer. Nevertheless, whether or not one approves of such gatherings, the story-telling times are invariably interesting, and much genuine information is passed around. Some of the rattlesnake hunters would seem not to know too much about the reptiles, but 'cowboys' like this are few since by its very nature the craft of rattlesnake hunting is not for the novice. Many of the hunters are extremely knowledgeable about the habits and behaviour of the animals they catch, and at the same time many myths that are as old as time are told again and perpetuated for the benefit of a new generation. The stories told sometimes cover a wider field than simply rattlesnakes, and one of the most absorbing topics of discussion is the number of deaths caused by snakebite around the world, and the reptiles that are responsible for them.

The World Health Organisation monitors records of human mortality due to snakebite. Of necessity the figures it receives cannot be regarded as absolutely accurate since in parts of the world where concentrations of venomous snakes occur, many lethal bites are never reported because of the problems of communications from distant parts of a country with no efficient telephone system, and where villagers cannot write. The figures therefore usually reflect those patients who have come or been brought to some place where they can receive treatment. This is often a hospital, but it may be some sort of clinic or field treatment centre. This does not give any indication of how many people are bitten by venomous snakes, and who subsequently recover from their injuries, as these will also very often go unrecorded away from the towns. For these reasons, estimates of snakebites are open to all sorts of interpretation, and one person's guess about the true figures is as likely to be as correct as the next person's. But for what they are worth, the WHO calculates that the number of bites from venomous snakes in India that result in the death of a human being is between 10,000 and 12,000 a year, which is about 0.005 per cent of the population. The figure for Burma is relatively higher at 2,000, which is 0.015 per cent of the population. The Ryukuk group of islands that lie between Japan and Taiwan provide the highest incidence of snakebite, 0.2 per cent of the population. Most of these are the result of encounters with the Habu, *Trimeresurus flavoviridis*, and although bites from this snake are singularly unpleasant the animal is not especially deadly, and as a result most of the victims survive.

In the whole of South America there are about 2,000 deaths a year, of which a third are the result of bites from a variety of species of *Bothrops*, and in Africa the Puff Adder, the Egyptian Cobra and the Saw-scaled Viper cause 1,000 deaths per annum. One might think there would be a very high incidence of death due to snakebite in Australia since the country is rich in venomous species, but virtually everyone there wears shoes, in contrast to places like India and Africa, and living in a developed nation patients are able to be whisked off to a hospital promptly in the majority of cases. As a result there are only 5 to 10 deaths a year, caused for the most part by the Tiger Snake and the Brown Snake. Throughout the whole of Europe there are only 15 deaths from snakebite per annum and most of these occur in the south-east of the continent and are caused by the Sand Viper.

The rattlesnake hunters would probably argue with the figure for the whole of the United States since it hardly sounds very dramatic, but the truth is that less than 15 people each year die from snakebite in the States, and most of these fatalities are the result of bites from the Eastern Diamondback Rattlesnake and the Western Diamondback Rattlesnake.

KILLER BEES

It would appear likely that the American irrational fear of rattlesnakes is likely to be matched in the not too distant future by a growing apprehension about the so-called Killer Bees from South America. Ten years ago it was thought that they would reach the United States in the early nineties. In fact, they have been advancing far more quickly than anyone anticipated at that time. In 1985 they arrived in Panama, and before long the country's honey industry had collapsed. Bee squads were formed by the government of that country, and in three years they had destroyed 27,000 colonies and swarms of Killer Bees. Beekeepers had to give up honey production when terrified neighbours attacked them, burnt their hives, and sued them for livestock that had been killed by the fierce little insects. The few that remain find beekeeping a far more expensive business than it used to be. At one time three men could easily handle a large number of hives. Today, that number has risen to eight, because extra staff are needed to be on hand at all times to keep the bees sedated with smoke while other men remove the honey from the hives.

In May 1985 a cargo ship from South America arrived in California with a swarm of Killer Bees. They were quickly destroyed, but it is worth bearing in mind that no country has yet managed to stop an invasion of these animals. Scientists who are studying the problem complain that in the United States no one is yet listening to their warnings, nor taking the threat seriously, but as one of them says, 'If you come across a quantity of Killer Bees and they start to attack you, you have only two choices, either run or die. It is as simple as that, running or dying.'

What to do about an invasion of Killer Bees is a difficult question to answer with time running out fast. It is now calculated that the insects will arrive, without the aid of ships, at the United States border sometime in 1989. One cannot rush out and destroy them wholesale as is thoughtlessly done with rattlesnakes, since pesticides, which would be the only way to do it, would also destroy existing stocks of honey bees, and honey bees are almost totally responsible for the pollination of food plants. Last year in the United States agriculture was worth $19 billion. True, the bees will not be able to travel too far north as the climate would not be too cold for them, but in the southern states they could cause untold harm, and almost certainly would provoke in people living in those areas the same sort of panic and hatred that rattlesnakes do already. The fact is, that though Killer Bees can attack and kill human beings, such incidents are not common. The danger is far greater to livestock, which is attacked more readily than man.

The whole situation is really difficult to resolve, since there is no doubt that Killer Bees are far better at producing honey than the conventional European Honey Bee, and in some parts of South America apiarists are learning to put up with the extra aggressive strain in the interests of honey production. In the States it has been suggested that public pressure might force the government to bring in legislation whereby licences would only be granted to beekeepers if they fulfilled a number of expensive husbandry measures to protect people and their livestock. If this happens, many of the 200,000 amateur beekeepers would be forced to give up their hobby. It takes a well equipped, professional set-up to maintain production with the new variety, which would inevitably join the new strain.

The layman asks the scientist for instant answers to the problem of aggression in honey bees, but as one of them pointed out, we do not understand aggression in our own species, and when we look at this little insect it is infinitely more difficult to even know where to start to study the phenomenon, and before we can control it we need to understand it. It will be interesting to see whether in a few years it is the bee that has become public enemy number one, instead of the rattlesnake. There is no doubt that man will always regard venomous animals with awe and fear but is always worth remembering that virtually all injuries and deaths that are caused by snakes, by spiders, and the rest of them, are almost invariably the direct result of our interference in the first place.

Some of that interference is unintentional, as when a villager in India accidentally treads on a sleeping snake, and in instances like this more care and education could reduce fatalities.

Venomous animals can be dangerous, but all of them would rather be left alone to get on with their own lives. They are fascinating and beautiful animals. Let us hope that they are still around in another generation. Their welfare and future lie in our hands.

Appendix: Suppliers of Antivenoms Worldwide

Antivenoms are manufactured by laboratories in a number of countries for use against their own venomous fauna. In this appendix the suppliers are listed alphabetically by country, also listed alphabetically. As far as is known, all the world's suppliers are included. The whole business of venom and antivenoms is a complex one, and suppliers should be contacted for details before using their products. In some cases a particular product may be of some use against species other than that for which it is intended. Such species are marked with an asterisk and if you are in the field, and desperate, it is worth trying them, but always check everything before you travel to parts of the world where venomous animals may be a definite hazard to your particular enterprise. Most manufactured antivenoms are designed for cases of snakebite; those that are not are marked with the preceding symbol =.

It is never safe to assume that all laboratories are producing antivenoms at a given moment since war and other disruptions may alter the situation, so it is advisable to contact relevant suppliers before departure to ascertain what is available.

COUNTRY	NAME OF PRODUCT	FOR USE AGAINST	
ALGERIA Institute Pasteur d'Algérie, Rue Docteur Laveran, Algiers, Algeria.	Antiviperin =Scorpion Antivenom	*Cerastes cerastes* *Vipera lebetina* *Androctonus australis*	Horned viper Levantine Viper
ARGENTINA Instituto Nacional de Microbiologia, Avenido Velez Sarsfield 563, Buenos Aires, Argentina.	Bothrops bivalent Tropical polyvalent Tropical trivalent	*Bothrops alternatus* *Bothrops neuwiedi* *Bothrops alternatus* *Bothrops jararaca* *Bothrops jararacussu* *Crotalus durissus terrificus* *Bothrops alternatus* *Bothrops neuwiedi* *Crotalus durissus terrificus*	Yarara Wied's Lance-head or Yarara Chica or Painted Jararaca Yarara Jararaca Yarara Cascabel Yarara Wied's Lance Head or Yarara Chica Cascabel
AUSTRALIA Commonwealth Serum Labs, 45 Poplar Road, Parkville, Victoria 3052, Australia	Death Adder Tiger Sea Snake Taipan Eastern Brown Snake Brown Snake Polyvalent (Australia/New Guinea) =Red-backed spider Antivenom =Sea wasp =Stonefish	*Acanthophis antarcticus* *Acanthophis pyrrhus* *Notechis scutatus* *Enhydrina schistosoa* **Austrelaps superba* **Pseudechis porphyriacus* **Tropidechis carinatus* *and at least 12 species of sea snake *Oxyuranus scutellatus* **Parademansis microlepidota* *Pseudonaja textilis* **Pseudonaja affinis* **Pseudonaja nuchalis* *Pseudechis australis* **Pseudechis porphyriacus* *Oxyuranus scutellatus* *Acanthophis antarcticus* *Notechis scutatus* *Pseudechis australis* *Pseudonaja textilis* **Austrelaps superba* **Pseudechis porphyriacus* **Pseudonaja affinis* **Pseudonaja nuchalis* **Pseudechis papuana* **Parademansis microlepidota* *Latrodectus mactans hasselti* *Chironex fleckeri* **Chiropsalmus quadrigatus* *Synanceja trachynis*	Common Death Adder Mainland Tiger Snake Beaked Sea Snake Taipan Eastern Brown Snake Mulga Common Death Adder Mainland Tiger Snake Mulga Eastern Brown Snake Red-backed spider Sea Wasp Stone Fish

COUNTRY	NAME OF PRODUCT	FOR USE AGAINST	
BRAZIL			
Instituto Butantan,	Anticrotalic	*Crotalus durissis*	Cascabel
Caixa Postal 65,		*terrificus*	
05504 Sao Paolo,	Antilaqetico	*Lachecis muta*	Bushmaster
Brazil.			Sururucu
	Antibothropico	*Bothrops jararaca*	Jararaca
		Bothrops moojeni	Moojen's Pit Viper
		Bothrops cotiara	Cotiara
		Bothrops alternatus	Urutu
		Bothrops jararacussu	Jararacussu
		Bothrops neuwiedi	Wied's Lance Head
	Antiophidico polyvalent	*Crotalus durissis*	Cascabel
		terrificus	
		Bothrops jararaca	Jararaca
		Bothrops moojeni	Hoojen's Pit Viper
		Bothrops cotiara	Cotiara
		Bothrops alternatus	Urutu
		Bothrops jararacussu	Jararacussu
		Bothrops neuwiedi	Wied's Lance Head
	Antibothropico-lachetico	*Lachesis muta*	Bushmaster
		Bothrops alternatus	Urutu
		Bothrops jararacussu	Jararacussu
		Bothrops jararaca	Jararaca
		Bothrops moojeni	Moojen's Pit Viper
		Bothrops cotiara	Cotiara
		Bothrops neuwiedi	Wied's Lance Head
	Antilapidico	*Micrurus fontalis*	Giant Coral Snake
		Micrurus corallinus	Veradeira
	=Antiarachnidico	*Phoneutria spp*	
	polyvalente	*Loxosceles spp*	
		Lycosa spp	
Syntex do Brasil S/A,	Crotalus	*Crotalus durissus*	Cascabel
Industria e Commercio,		*terrificus*	
Caixa Postal 951,	Bothrops	*Bothrops alternatus*	Urutu
Sap Paolo, Brazil.		*Bothrops atrox asper*	Barba Amarilla
		Bothrops jararaca	Jararaca
		Bothrops jararacussu	Jararacussu
		Bothrops cotiara	Cotiara
BULGARIA			
Institute of Epidemiology	Viper	*Vipera ammodytes*	Long-nosed viper
and Microbiology,		*Vipera berus*	European Viper
Sofia, Bulgaria.		*Vipera aspis*	Jura Viper
BURMA			
Industrial and	Siamese Cobra	*Naja naja kaouthia*	Siamese Cobra
Pharmeceutical Corporation,	Russell's Viper	*Vipera Russelli*	Russell's Viper
Rangoon, Burma.		*siamensis*	
	Bivalent	*Naja naja kaouthia*	Yellow Cobra
		Vipera russelli	Russell's Viper or Tic
			Polonga or Daboila
CHINA			
Shanghai Vaccine and	Mamushi monovalent	*Agkistrodon halys*	Mamushi
Serum Institute,	Monovalent	*Agkistrodon acutus*	100 Pace Snake
1262 Yang An Road (W),			
Shanghai, China.			

COUNTRY	NAME OF PRODUCT	FOR USE AGAINST	
COLOMBIA			
Instituto Nacional de Salud,	Antiophidico	*Bothrops atrox asper*	Barba Amarilla
Avenida Eldorado con		**Bothrops spp*	
Carrera,	Polyvalent	*Crotalus durissus*	Cascabel
Zona G.		*terrificus*	
Bogota, Colombia.		**Crotalus spp*	
COSTA RICA			
Universidad de Costa Rica,	Antilaquesico	*Lachesis muta*	Bushmaster
Cuidad Universitaria		*stenophrys*	
Rodrigo Facio,		**Lachesis muta muta*	Bushmaster
San Jose, Costa Rica.		**Lachesis muta noctiyaga*	Bushmaster
	Polyvalent	*Bothrops atrox asper*	Terciopelo
		Crotalus durissus	Central American
		terrificus	Rattlesnake
		Lachesis muta stenophrys	Bushmaster
		*Lachesis muta muta	Bushmaster
		*Lachesis muta noctiyaga	Bushmaster
		*Agkistrodon bilineatus	
		*Bothrops nummifer	
		*Bothrops picadoi	
		*Bothrops nasutus	
		*Bothrops ophryomegas	
		*Bothrops godmani	
		*Bothrops lateralis	
		*Bothrops schlegeli	
		*Bothrops nigroviridis	
	Anti-coral	*Micrurus nigrocinctus*	Coral Snake
		nigrocinctus	
		Micrurus nigrocinctus	Coral Snake
		mosquitensis	
		*Micrurus carinicaudus	Coral Snake
		dumerili	
		*Micrurus fulvius fulvius	Coral Snake
CZECHOSLOVAKIA			
Institute for Sera and	Venise	*Vipera ammodytes*	Long-nosed Viper
Vaccines,		*Vipera berus*	European Viper
West Pieck Str,			
Prague, Czechoslovakia.			
ECUADOR			
Instituto Nacional de	Antibothrops	*Bothrops atrox asper*	Barba Amarilla
Higiene Guayaquil,			
Ecuador.			
ENGLAND			
Lister Institute of	=Scorpion	*Androctonus australis*	
Preventative Medicine,		*Buthus occitanus*	
Elstree,		*Leiurus quinquestriatus*	
Herts WD6 3AX, England.			

COUNTRY	NAME OF PRODUCT	FOR USE AGAINST	
FRANCE			
Institut Pasteur,	Ipser V	Vipera aspis	Jura Viper
Annexe de Garches,		Vipera berus	European Viper
92 Hauts-de-Seine,	Ipser Europe	Vipera ammodytes	Long-nosed Viper
Paris, France.		Vipera aspis	Jura Viper
		Vipera berus	European Viper
	Bitis–Echis–Naja	Bitis arietans	Puff Adder
		Bitis gabonica	Gaboon Viper
		Bitis nasicornis	Rhinoceros Viper
		Echis carinatus	Saw-scaled Viper
		Haemachatus haemachatus	Ringhals
		Naja haje	Egyptian Cobra
		Naja melanoleuca	Forest Cobra
		Naja nigricollis	Spitting Cobra
		Naja nivea	Cape Cobra
	Near and Middle East	Vipera ammodytes	Long-nosed Viper
		Vipera lebetina obtusa	Levantine Viper
		Vipera xanthina palestinae	Palestine Viper
		Cerastes cornutus	Horned Viper
		Cerastes vipera	Avicenna's Viper
		Echis carinatus	Saw-scaled Viper
		Naja naja	Indian Cobra
		Naja haje	Egyptian Cobra
	Cobra	Naja naja kaouthia	Yellow Cobra
	Dendroaspis	Dendroaspis angusticeps	Eastern Green Mamba
		Dendroaspis jamesoni	Jameson's Mamba
		Dendroaspis polylepis	Black Mamba
		Dendroaspis viridis	Western Green Mamba
GERMANY			
Behringwerke AG,	Europe	Vipera ammodytes	Long-nosed Viper
D3550 Marburg/Lahn,		Vipera berus	European Viper
West Germany.		*Vipera aspis	Jura Viper
		*Vipera lebetina	Levantine Viper
		*Vipera xanthina	Palestine Viper
	North Africa	Bitis gabonica	Gaboon Viper
		Echis carinatus	Saw-scaled Viper
		Naja haje	Egyptian Cobra
		Vipera lebetina	Levantine Viper
		*Cerastes cerastes	Horned Viper
		*Cerastes vipera	
		*Bitis arietans	Puff Adder
		*Naja melanolueca	Forest Cobra
		*Naja nigricollis	Black-necked Cobra
	Central Africa	Bitis gabonica	Gaboon Viper
		Dendroaspis polylepis	Black Mamba
		Naja haje	Egyptian Viper
		*Bitis arietans	Puff Adder
		*Bitis nasicornis	Rhinoceros Viper
		*Dendroaspis viridis	Green Mamba
		*Haemachatus haemachatus	Ringhals
		*Naja melanoleuca	Forest Cobra
		*Naja nigricollis	Spitting Cobra
	Near and Middle East	Echis carinatus	Saw-scaled Viper
		Naja haje	Egyptian Cobra
		Vipera ammodytes	Long-nosed Viper
		Vipera lebetina	Levantine Viper
		*Cerastes cerastes	Horned Viper
		*Vipera xanthina	Palestine Viper
		*Cerastes cornutus	Horned Viper

COUNTRY	NAME OF PRODUCT	FOR USE AGAINST	
INDIA			
Central Research Institute, Kasauli, Simla Hills, HP, India.	Polyvalent	*Bungarus caeruleus*	Indian Krait
		Naja naja	Indian Cobra
		Vipera russselli	Russell's Viper
		Echis carinatus	Saw-scaled Viper
		*Bungarus fasciatus	Banded Krait
		*Ophiophagus hannah	King Cobra
Haffkine Bio-pharmaceutical Corporation Ltd., Parel, Bombay, India.	Bungarus	*Bungarus caeruleus*	Common Krait
		*Bungarus fasciatus	Banded Krait
	Naja	*Naja naja*	Indian Cobra
		*Naja naja kaouthia	
	Vipera	*Vipera russelli*	Russell's Viper or Tic Polonga or Daboia
		Naja naja oxiana	
	Echis	*Echis carinatus*	Saw-scaled Viper
		*Ophiophagus hannah	King Cobra
	Polyvalent	*Bungarus caeruleus*	Common Krait
		Naja naja	Indian Cobra
		Echis carinatus	Saw-scaled Viper
		Vipera russelli	Russell's Viper or Tic Polonga or Daboia
		*Trimeresurus gramineus	Green Pit Viper
		*Trimeresurus labialis	
INDONESIA			
Perusahaan Negara BioFarma, 9 Jalan Pasteur, Bandung, Indonesia.	Agkistrodon	*Agkistrodon rhodostoma*	Malayan Pit Viper
	Bungarus	*Bungarus fasciatus*	Banded Krait
	Naja	*Naja sputatrix*	Malayan Cobra
IRAN			
Institut d'Etat des Serums et Vaccines Razi, PO Box 656, Teheran, Iran.	Naja	*Naja naja oxiana*	Oxus Cobra
	Vipera	*Vipera lebetina*	Levantine Viper
	Echis	*Echis carinatus*	Saw-scaled Viper
	Pseudocerastes	*Pseudocerastes persicus*	Persian Horned Viper
	Viper	*Vipera latasti*	Snub-nosed Viper
	Agkistrodon	*Agkistrodon halys*	Mamushi
	Polyvalent	*Naja naja oxiana*	Common Cobra
		Vipera lebetina	Levantine Viper
		Vipera xanthina	Palestine Viper
		Echis carinatus	Saw-scaled Viper
		Pseudocerastes persicus	
		Agkistrodon halys	
		*Cerastes cerastes	Horned Viper
		*Eristocophis macmahoni	
		*Vipera aspis	Jura Viper
		*Vipera cerastes	
		*Vipera latasti	
		*Vipera xanthina palestinae	Palestine Viper
	=Polyvalent scorpion Serum	*Andoctonus crassicauda*	
		Buthotus saulcyi	
		Hemiscorpius lepturus	
		Mesobuthus eupeus	
		Odontobuthus doriae	
		Scorpio maurus	

COUNTRY	NAME OF PRODUCT	FOR USE AGAINST	
ISRAEL			
Rogoff Medical Research Institute,	Arabian Echis	*Echis coloratus*	Arabian Saw-scaled Viper
Beilinson Medical Centre, Tel Aviv, Israel.	Palestine Viper	*Vipera xanthina Palestinae*	Palestine Viper
ITALY			
Instituto Sieroterapico e Vaccinogeno Toscano Sclavo, Via Florentina 1, Siena, Italy.	Antiviperin	*Vipera ammodytes* *Vipera aspis* *Vipera berus* *Vipera ursini*	Long-nosed Viper Jura Viper European Viper Ursini's Viper
JAPAN			
The Chemo-Sero-Therapeutic Research Institute Kunamoto 860, Kyushu, Japan.	Habu Antivenine	*Trimeresurus flavoviridis*	Habu
The Takeda Pharmaceutical Company, Osaka, Japan.	Mamushi Antivenine	*Agkistrodon halys*	Mamushi
Research Institute for for Microbial Diseases, Osaka University, Suite 565, Osaka, Japan.	Mamushi Antivenine	*Agkistrodon halys*	Mamushi
Kiteasato Institute, Minato-ku, Tokyo, Japan.	Mamushi Antivenine	*Agkistrodon halys*	Mamushi
Chiba Prefectural Serum Institute, Inchikawa, Japan.	Mamushi Antivenine	*Agkistrodon halys*	Mamushi
MEXICO			
Laboratorios MYN S/A, Avenida Coyoacan 1707, Mexico City 12 DF, Mexico.	Monovalent bothrops Polyvalent crotalus	*Bothrops atrox asper* *Crotalus atrox*	Barba Amarilla Western Diamondback Rattlesnake
		Crotalus durissus terrificus	Cascabel
		Crotalus tigris *All native crotalids	Tiger Rattlesnake
	Polyvalent Mexico	*Bothrops atrox asper* *Crotalus durissus terrificus*	Barba Amarilla Cascabel
		Crotalus tigris *Crotalus atrox*	Tiger Rattlesnake Western Diamond Rattlesnake
	=Antialacrás polyvalent	*Centruroides suffusus* *Centruroides noxius* *Centruroides limidus*	

COUNTRY	NAME OF PRODUCT	FOR USE AGAINST	
Instituto Nacional de Higiene Avenida M. Escobedo 20, Mexico City DF, Mexico.	Antibothrops	Bothrops atrox asper	
	Anticrotalus	Crotalus basiliscus basiliscus	Mexican Rattlesnake
		Crotalus durissus terrificus	Cascabel
	Polyvalent	Bothrops atrox asper	Western Diamond Rattlesnake
		Crotalus basiliscus basiliscus	
		Crotalus durissus terrificus	Cascabel
	=Antialacrás polyvalent	Centruroides noxius	
Laboratorio Zapata, Mexico City DF, Mexico	=Antialacrás polyvalent	Centruroides suffusus Centruroides noxius	

PERU

COUNTRY	NAME OF PRODUCT	FOR USE AGAINST	
Instituto Nacional de Higiene Lima, Peru.	Bothrops polyvalent	Bothrops atrox asper	Barba Armarilla
		Lachesia muta	Bushmaster
		*Bothrops spp	
	Anticoral polyvalent	Micrurus nigrocinctus	Giant Coral Snake
		Micrurus mipartitus	Cobra Coral Snake
		Micrurus frontalis	Coral Snake
		*Micrurus fulvius fulvius	Coral Snake
		*Micrurus alleni	Coral Snake
		*Micrurus carinicaudus	Coral Snake
		*Micrurus spixi	Coral Snake
		*Micrurus lemniscatus	Coral Snake
		*Micrurus corallinus	Coral Snake
	=Anti-loxoscelico	Loxosceles spp	

PHILIPPINES

COUNTRY	NAME OF PRODUCT	FOR USE AGAINST	
Serum and Vaccine Labs, Alabang, Mutinlupa, Rizal, Philippines.	Cobra	Naja naja philippinensis	Philippine Cobra

RUSSIA

COUNTRY	NAME OF PRODUCT	FOR USE AGAINST	
Research Institute of Vaccine and Serum, Ministry of Public Health, UI Kafanova 93, Tashkent, USSR.	Monovalent echis carinatus	Echis carinatus	Saw-scaled Viper
	Monovalent naja naja	Naja naja	Indian Cobra
	Monovalent vipera lebetina	Vipera lebetina	Levantine Viper
	Polyvalent naja and echis	Echis carinatus Naja naja	Saw-scaled Viper Indian Cobra
	Polyvalent vipera and naja	Naja naja Vipera lebetina	Indian Cobra Levantine Viper

SOUTH AFRICA

COUNTRY	NAME OF PRODUCT	FOR USE AGAINST	
South African Institute for Medical Research, PO Box 1038, Johannesburg 2000, Republic of South Africa	Polyvalent	Haemachatus haemachatus	Ringhals
		Naja nivea	Cape Cobra
		Naja haje	Egyptian Cobra
		Naja melanoleuca	Forest Cobra
		Naja nigricollis	Spitting Cobra

COUNTRY	NAME OF PRODUCT		FOR USE AGAINST
South African Institute for Medical Research, Johannesburg, Republic of South Africa.		*Dendroaspis angusticeps*	Eastern Green Mamba
		Dendroaspis jamesoni	Jameson's Mamba
		Dendroaspis polylepis	Black Mamba
		Bitis arietans	Puff Adder
		Bitis gabonica	Gaboon Viper
		*Naja naja	Indian Cobra
		*Ophiophagus hannah	King Cobra
		*Pseudohaje goldi	
		*Walterinnesia egyptia	
		*Dendroaspis viridis	Green Mamba
	Echis	*Echis carinatus*	Saw-scaled Viper
		*Echis coloratus	
		*Cerastes cerastes	Horned Viper
		*Cerastes vipera	
	Boomslang antivenom	*Dispholidus typus*	Boomslang
	=Black Widow	*Latrodectus mactans*	Black Widow
	=Scorpion	*Parabuthus spp*	

TAIWAN

COUNTRY	NAME OF PRODUCT		FOR USE AGAINST
National Institute of Preventive Medicine, 16 Kun Yang Street, Nan Kang, Taipei, Taiwan.	Agkistrodon	*Agkistrodon acutus*	Long-nosed Pit Viper
		*Trimeresurus microsquamatus	
	Bungarus	*Bungarus multicinctus*	Many Banded Krait
	Naja	*Naja naja atra*	Chinese Cobra
	Trimeresurus	*Trimeresurus stejnegeri*	Bamboo Viper
		Trimeresurus microsquamatus	Chinese Habu
		*Agkistrodon acutus	
	Naja–Bungarus	*Bungarus multicinctus*	
		Naja naja atra	Common Cobra

THAILAND

COUNTRY	NAME OF PRODUCT		FOR USE AGAINST
Queen Saovabha Memorial Institute, Rama 4 Road, Bangkok, Thailand	Bungarus	*Bungarus fasciatus*	Banded Krait
	Cobra	*Naja naja*	Indian Cobra
	King Cobra	*Ophiophagus hannah*	King Cobra
	Russell's Viper	*Vipera russelli*	Russell's Viper
	Malayan Pit Viper	*Agkistrodon rhodostoma*	Malayan Pit Viper
	Green Tree Viper	*Trimeresurus albolabris*	Green Tree Viper
		Trimeresurus erythrurus	

UNITED STATES OF AMERICA

COUNTRY	NAME OF PRODUCT		FOR USE AGAINST
Wyeth Laboratories, Box 8299, Philadelphia, Pennsylvania, USA.	Antivenin (crotalidae)	*Crotalus durissus terrificus*	South American Rattlesnake
	Polyvalent	*Bothrops atrox asper*	Barba Amarilla
		Crotalus adamanteus	Eastern Diamond Rattlesnake
		Crotalus atrox	Western Diamond Rattlesnake
		*Crotalus spp	
		*Sisturus spp	
		*Agkistrodon spp	(Old and New World)
		*Bothrops spp	
		*Lachesis spp	
		*Trimeresurus spp	
	Antivenin (Micrurus fulvius)	*Micrurus fulvius fulvius*	Eastern Coral Snake
		*Micrurus fulvius tenere	

COUNTRY	NAME OF PRODUCT	FOR USE AGAINST	
Merck, Sharp and Dome, Westpoint, Pennsylvania 19486, USA	=Black Widow	*Latrodectus mactans*	Black Widow

VENEZUELA

COUNTRY	NAME OF PRODUCT	FOR USE AGAINST	
Laboratorio Behrens, Avenida Principal de Chapellin, Apartado 62, Caracas 101, Venezuela.	Crotalus	*Crotalus durissus terrificus* *Crotalus vegrandis*	Cascabel
	Bothrops	*Bothrops atrox asper* *Bothrops venezuelae* *Bothrops atrox asper* *Bothrops columbiensis*	Barba Armilla Tigra mariposa Barba Amarilla
	Bothrops/crotalus	*Bothrops venezuelae* *Crotalus durissus terrificus* *Bothrops columbiensis* *Bothrops bilineata* *Bothrops lansbergi* *Bothrops lichenosus* *Bothrops medusa* *Bothrops neglectus* *Bothrops schlegeli* *Crotalus vegrandis*	Tigra mariposa Cascabel

YUGOSLAVIA

COUNTRY	NAME OF PRODUCT	FOR USE AGAINST	
Institute of Immunology, Rockefellerova 2, Zagreb, Yugoslavia.	Antiviperinum	*Vipera ammodytes* *Vipera berus* *Vipera aspis*	Long-nosed Viper European Viper Jura Viper

Bibliography

Arnold, Robert E. *What To Do About Bites and Stings Of Venomous Animals* (Collier Books, 1973)

Ash, Russell and Lake, Brian. *Frog Raising For Pleasure And Profit, And Other Bizarre Books* (Macmillan, 1985)

Author not credited. *Watch Out: Surviving Venomous Bites* (Hospital and Benefits Association, Australia, 1983)

Ballasina, D. *Amphibians Of Europe* (David & Charles, 1984)

Barzdo, John (ed). *Traffic Bulletin*, various issues (Wildlife Trade Monitoring Unit)

Behler, John and King, Wayne. *The Audubon Society Field Guide To North American Reptiles And Amphibians* (Alfred A. Knopf, 1979)

Bucherl, W. and Buckley, E. *Venomous Animals And Their Venoms* (Academic Press, 1971)

Burland, Cottie. *North American Indian Mythology* (Newnes Books, 1986; Peter Bedrick Books, 1985)

Burton, Robert. *Venomous Animals* (Crescent Books, 1975)

Cable, B. et al. *Prolonged Defibrination After A Bite From A Non-Venomous Snake (Rhabdophis subminatus)* (Journal of the American Medical Association, 251, 1984)

Caras, Robert A. *Venomous Animals Of The World* (Prentice Hall, 1974)

— *Dangerous To Man* (Rinehart & Winston, 1975)

Coborn, John. *Snakes and Lizards* (David & Charles, 1987)

Cloudesley-Thompson, D. L. *Spiders and Scorpions* (Bodley Head, 1973)

Cochran, Doris. *Living Amphibians Of The World* (Hamish Hamilton, 1961)

Corbet, Gordon and Ovenden, Denys. *The Mammals of Britain And Europe* (Collins, 1980)

Cox, Graham. *Tropical Marine Aquaria* (Hamlyn, 1971)

Craven, Roy. *A Concise History Of Indian Art* (Thames & Hudson, 1976)

Davis, Wade. *The Serpent And The Rainbow* (Fontana, 1987; Warner Books, Inc; 1982)

Dodge, Natt. N. *The Poisonous Dwellers Of The Desert* (Southwest Park and Monuments Association, USA, 1974)

Dunson, W.A. *The Biology Of Sea Snakes* (University Press, 1975)

Ellis, Michael D. *Dangerous Plants, Snakes And Marine Life* (Drug Intelligence Publications Inc, USA, 1974)

Ferns, Ann. *Watch Out! These Creatures Bite And Sting* (RPLA, Australia, 1983)

Fichter, George. *Insect Pests* (Hamlyn; Western Publishing Co. Inc, 1966)

Fitchett, Ralph (ed). *Cobwebs,* various issues (Tropical Butterfly Gardens, Cleethorpes)

Fleming, Ian. *Doctor No* (Jonathan Cape, 1958; Berk, 1985)

Ford, R.L.E. *Studying Insects* (Frederick Warne, 1973)

Gardner, John. *Amber Nine* (Corgi, 1967)

Garnet, J. Ros. *Venomous Animals Dangerous To Man* (Commonwealth Serum Laboratories Publications, Australia, 1968)

Gifford, Dennis. *The Complete Catalogue Of British Comics* (Webb & Bower, 1985; Viking, 1987)

Ginsburg, C.M. *Fire Ant Envenomation In Children* (Pediatrics, 73, 1984)

Halstead, Bruce W. *Dangerous Marine Animals* (Cornell Maritime Press, 1959; Cornell Maritime Press, Inc, 1980)

Hardy, Phil (ed). *The Aurum Film Encyclopedia Of Horror* (Aurum Press, 1985; Harper & Row, 1987)

Harmon, R.W. and Pollard, C.B. *Bibliography Of Animal Venoms* (University of Florida Press, 1948)

Horsman, Paul. *The Seafarer's Guide To Marine Life* (Croom Helm, 1985)

International Society of Toxinology. *Newsletter,* various issues

Ions, Veronica. *Egyptian Mythology* (Newnes Books, 1986; Peter Bedrick Books, 1983)

Jackman, L.A.J. *Sea Water Aquaria* (David & Charles, 1974)

Klein, Dan, McLelland, Nancy and Haslam, Malcolm. *In The Art Deco Style* (Thames & Hudson, 1987; Rizzoli, 1986)

La Barre, Weston. *They Shall Take Up Serpents* (University of Minnesota Press, 1962)

Lerker, Manfred. *The Gods And Symbols Of Ancient Egypt* (Thames & Hudson, 1974)

Losito, Linda. *Amphibians and Reptiles* (Facts on File, 1989)

Mattison, Chris. *Frogs & Toads of the World* (Facts on File, 1987)

   — *Snakes of the World* (Blandford Press, 1987; Facts on File, 1987)

McWhirter, Norris D. (ed). *The Guinness Book of Records* (Guinness Books, 1984)

— (ed). *The Guinness Book of Animal Facts and Feats* (Guinness Books, 1986)

Minton, Sherman A. *Venom Diseases* (Charles C. Thomas, 1974)

Minton, Sherman and Minton, Madge Rutherford. *Venomous Reptiles* (Allen & Unwin, 1971)

Morris, Ramona and Desmond. *Man And Snakes* (McGraw Hill, 1965)

National Geographic Society. *National Geographic Magazine*, many issues

Neubert, Otto. *The Story Of Tutenkhamen* (Dragon Books, 1972)

Nichol, John. *The Animal Smugglers* (Christopher Helm, 1987; Facts on File, 1987)

— *The Complete Guide To Pet Care* (Christopher Helm, 1988)

O'Toole, Christopher and Losito, Linda. *The Holiday Naturalist In Italy* (Christopher Helm, 1987; Stephen Greene Press, 1987)

Pickwell, G.V. And Evans, W.E. (eds). *The Handbook Of Dangerous Animals For Field Personnel* (Undersea Surveillance and Ocean Sciences Department, USA, 1972)

Putney, Susan K. and Wrightson, Berni. *The Amazing Spiderman* (Marvel Comics Group, 1986)

Russell, F.E. and Gertsch, W.J. 'Arthropod Bites' (Toxicon 21, 1983)

Russell, Findlay E. and Saunders, Paul R. *Animal Toxins* (Pergamon Press, 1967)

Schmidt, Karl and Inger, Robert. *Living Reptiles Of The World* (Hamish Hamilton, 1957)

Smith, Andrew M. *Tarantula Classification And Identification Guide* (Fitzgerald Publications, 1987)

Stahnke, Herbert L. *The Treatment Of Venomous Bites And Stings* (Arizona State University, 1966)

Webb, Ann (ed). *Journal of the British Tarantula Society,* many issues

— *Wall to Wall Spiders* (Imprint Books, 1987)

Webb, Ann and Frank. *Breeding Live Food For Reptiles And Tarantulas* (Fitzgerald, 1987)

Wheeler, Alwynne. *The World Encyclopedia Of Fishes* (MacDonald, 1985)

Wood, Gerald L. *The Guinness Book Of Pet Records* (Guinness Books, 1983)

World Health Organisation. *Progress In The Characterisation Of Venoms and Standardisation Of Antivenoms*, Offset Publications No 58 (World Health Organisation, Geneva, 1981)

Zilio, Marc. *Dangerous Animals Of The Sea* (Transworld, 1977)

Index